The Cooperating Universe

The Cooperating Universe

HOW NATURE CREATES COMPLEXITY

• • •

Dr. Werner J. Krieglstein &
Daniel Krieglstein M.S.

ISBN-13: 9781533330611
ISBN-10: 1533330611
Library of Congress Control Number: 2016908808
CreateSpace Independent Publishing Platform
North Charleston, South Carolina

Table of Contents

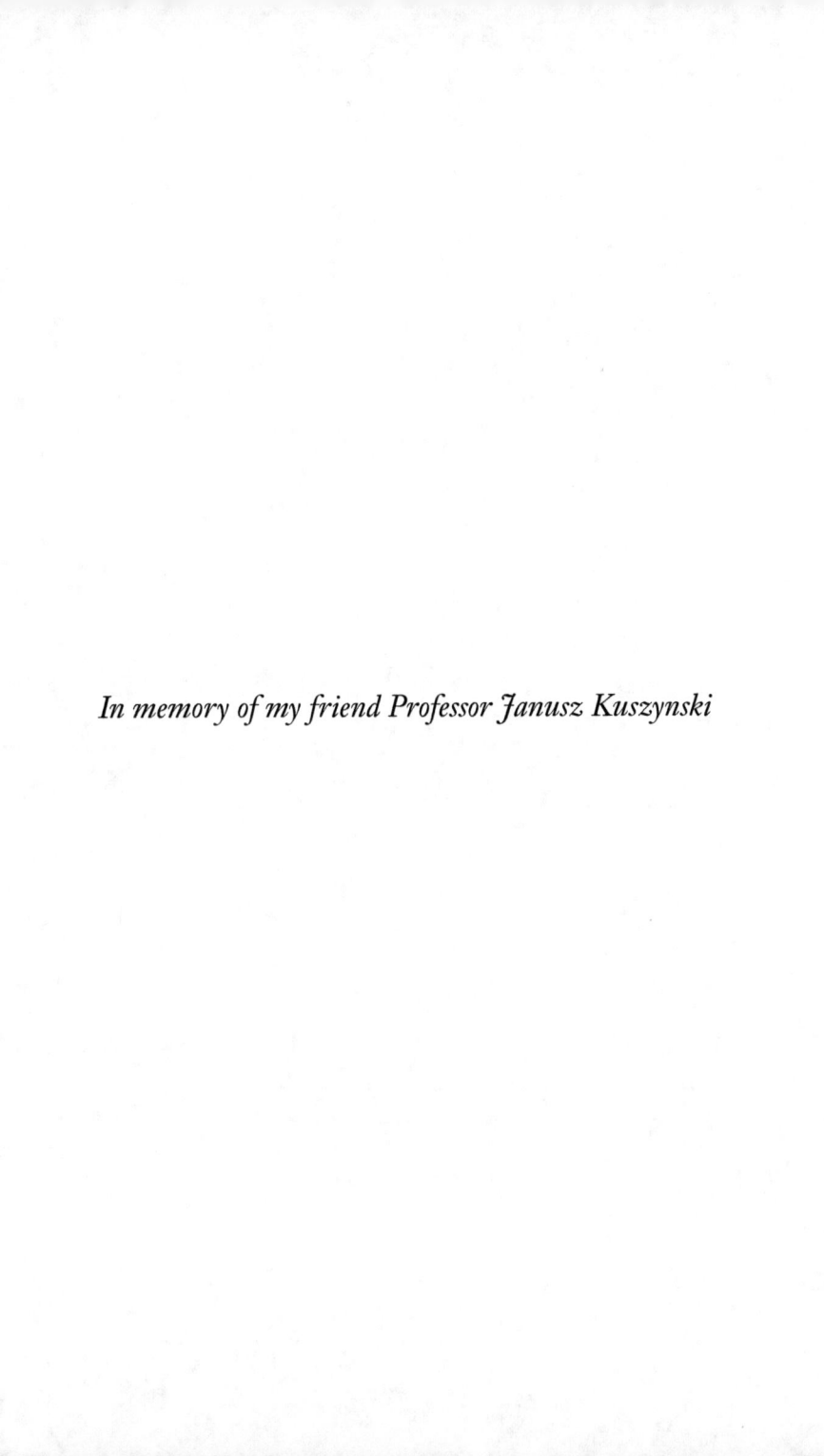

In memory of my friend Professor Janusz Kuszynski

Preface

● ● ●

THIS WORK IS THE RESULT of a life of learning and pondering. It would not have been possible without the consistent help, patience, and understanding of the many people I have encountered. Above all, there is my family. My five sons have contributed richly through deep conversations, lively discussions, and critical insights. Each is a professional in his own right, but none has ever grown tired of sharing his wisdom with me. Whenever they came across a book, an article, or a new thought, they passed it on to me, thus helping me keep up with the most current publications and materials across a variety of fields. This was especially important because the discussed topic addresses so much. In writing a book about the cooperating universe, my own family became a beautiful example of cooperation. From my son Robin I learned about advances in positive psychology; my son Mark, a filmmaker, filled me in on the newest discoveries in astronomy; Tom taught me to trust in crowd sourcing;

and Michael let me in on the deep intricacies of music. I still reference Michael's paper on entropy and the big bang, which he prepared for a symposium in Dubrovnik years ago. Then there is Daniel. Daniel, a psychologist, never gave up on me. He helped me through every aspect of the book and contributed major parts to the writing, correcting, and clarifying. Last but not the least, I must thank Maryann, my partner for nearly fifty years. As a social worker, she has never let me forget a passion for humanity and practical sense of caring.

Finally, I would like to thank my students. Their involvement gave me the courage to move along. Of my teachers, I must single out two professors who were my mentors. The first is Theodor W. Adorno. As a student at the Goethe University in Frankfurt, I was inspired by Professor Adorno to think big and never to shy away from complex thoughts. Next is my teacher Walter Gerstberger who, by example, taught me never to give up searching for the deepest secrets of the universe.

Introduction

● ● ●

OUR UNIVERSE IS NOT THE cold and uncaring place once described by Newtonian physics. That was a story about a perfectly tuned machine. In Newton's universe everyone's fate was already determined and little changed as a result of your existence. Human life, indeed, was utterly independent from the gears that moved much of this clock-like universe. Though the equations and experiments that gave birth to this narrative have undergone a fundamental change, this is the story we still often hear today when scientists describe the nature of our universe. One of the last big paradigm shifts happened when the equations of quantum physics and chaos theory provided a glimpse into a world vastly different from the familiar one we experience. That change was so great it altered the very questions we ask. We learned that the universe around us was intricately more connected than we could have previously imagined, even across vast distances of space.

Despite these foundational changes to the equations we use to describe nature, most people still think science is telling us the same familiar story about the nature of our universe. That the vast expanses of stars and planets have little in common with human existence; only now physicists generally agree that the our universe has a chaotic and probabilistic character. There have been several attempts to use the equations of chaos theory and quantum physics to argue for other paradigm shifts. A common fallacy among these attempts has been the desire to attribute the weird workings of quantum particles to the physically larger world that humans experience. Where scientists see weird connections between particles over extremely vast distances, the advocates for some of these attempted new paradigms argued that this meant there must also be weird connections between human experiences across vast distances. But the truth is, scientific experiments in quantum physics neither demonstrate this human level connection, nor predict it. These types of pop culture beliefs have made many scientists skeptical about any claims of a new scientific paradigm for the nature of our universe, and rightly so.

Today a new paradigm appears on the horizon. Our newly emerging paradigm needs to address such skepticism. Here we find that our paradigm's strength comes from the fact that it is a story the scientists themselves are telling, and relies wholly on the experimental discoveries of our universe as we currently know it. This new paradigm is marked by a new understanding of cooperation.

In the past two decades the debate over the origin of cooperative behavior, long believed to be an aberration, has steadily increased. A number of publications are now dealing directly with the topic.[1] A recent report on National Public Radio stated: "There are researchers in economics, evolutionary biology and psychology, and the law all training their own lenses on this multidisciplinary field that has become known as cooperation theory."

The story of cooperation theory begins with an understanding of energy and the role it plays in the lifespan of all things. When we examine the world around us, we can easily understand the basic role different types of energies play in everyday life. Our cellular phones need electrical energy to work. When the battery runs low, we plug them into a source of electrical energy to charge the battery. Most of our cars run on combustion energy. Some of us may not understand what that means, but we know that gas goes in and the car keeps moving when we turn the key. If the gas tank goes empty, the car stops moving. Our bodies run on biological energy. Even fewer of us understand how food is turned into muscle movements, but we all feel hunger and know we need food to survive. These examples all share the same pattern, of energy going into

---

1 Yochai Benkler, *How Cooperation Triumphs over Self Interest*, Dacher Keltner *"born to be good - the science of a meaningful life."*
S. Bowles and H. Gintis *A Cooperative Species, Human Reciprocity and Its Evolution*
Sterelny, Kim, Joyce, Richards, Calcott, Brett and Frazer, *Cooperation and Its Evolution*.

a system in order to keep it working in various ways. Each example consumes a different type of energy: electrical, combustion, and biological.

Our hunger story stands out from the other two examples in that if we don't eat, if we run out of energy, we will die. Not only will we die, but the cells of our body will slowly start to break down into dust. In contrast, when my car runs out of gas, I can leave it empty for a month and in all likelihood, the engine will still start when I turn the key. If I leave that gas tank empty for a few years, I'll need to clean some parts, but with the right storage the engine can still produce combustion energy from gasoline. However, when we talk to a particle physicist about our car, she will tell a story that closely resembles the energy needs of our body. It's just that this takes a much longer time to play out. More importantly, it plays out on a scale much too small for the human eye. On the microscopic level we find parallels between the decaying human body and the rusting car door. The story she tells sounds familiar because the small particles that make up our car also need energy to survive.

Let's focus on the car door's aluminum metal frame. From our perspective, the aluminum panel is a solid piece of metal, strong enough to protect us from small accidents. For the particle physicist, that aluminum door is just a series of particles being held together by various forms of energy. From the larger molecules to the smaller quarks, these particles require energy to bind together and survive as a unit, just like our body. Our physicist might even say the particles themselves are energy, bound together. The tiny size of these particles means the energy they need is a very small quantity. However, if new energy is not added to the system, the bonds that hold these particles together will eventually fail. Given enough time, our aluminum car door will slowly start to break apart. This example is a small introduction to the scientific concept of entropy. We will explore entropy in more depth, but for now it is important to know that the entire universe is struggling to overcome entropy. From the largest star to the smallest particle, we share a need for useable forms of energy. This struggle against entropy is the foundation of our new paradigm.

One final thought before we continue with our story. Philosophy and science have recently been portrayed as existing in a state of competition for truth. Some prominent scientists even suggest that philosophy is no longer needed, as if data and repeatability are all that is required in our pursuit of truths about the universe we live in. If you find yourself empathizing with this position, or are simply curious about the relationship between philosophy and science, you are encouraged to conduct a modest literature review of the "Philosophy of Science". Much of the interpretation of scientific experiments, the theories we use to make scientific predictions, are based on which scientific philosophy we adhere to. For instance, depending on the philosophical position of the scientists interpreting the data, our repeatable studies are either describing a truth about realty, or simply falsifying provably wrong explanations for reality but never actually reaching "truth". I bring this up now because the story being told here is one of both philosophy and science. There are philosophical implications to understanding what the current science is telling us about the nature of our universe, and the connections between us and the world we inhabit.

In a public forum I attended, the famous physicist Leon Lederman was asked to elaborate on the relationship between philosophy and science. He described science like a pyramid. Physics constituted the solid base, while chemistry occupied the softer middle, and the life sciences were on the top. And philosophy? She stands on the outside jealously gazing at the beautiful structure.

CHAPTER 1

E Pluribus Unum—Out of Many, One

• • •

EVER SINCE SOMETHING CAME FROM nothing—or nothing became something—everything is in a constant struggle to overcome death. Scientifically, this is expressed by the second law of thermodynamics. This empirically validated law of physics states that all things, left to their own devices, will fall into ever-greater disorder. The ultimate outcome of this general trend toward chaos is death—the return of something back to nothing; no plants or animals, no planets, no stars, no galaxies, no movement or variation at all from one place to another. Just an endless and uniform field of nothingness. What physicist call the heat death of our universe.

Lucky for us, the second law of thermodynamics is preceded by the first law, which states that all energy, in the final account, remains constant. Energy must be preserved. This means that ultimately, should everything return to nothing; nothingness is in confrontation with all the energy there is. Pure energy, the sum of all matter in the universe, is in constant struggle with nothingness; or, to express it somewhat differently, there is pure energy, and besides pure energy, there is nothing.

This leads us to some of the biggest unsolved questions in modern science: If the second law reigns supreme, how is it possible for life to escape its throngs, even if only temporarily? How can order exist in a sea of chaos? How can conscious beings, such as you and I, evolve from an otherwise completely unintentional universe? How could mind emerge from mindless matter? How could anything complex emerge from this impending demise?

It appears that to prevent this "falling back into nothingness", the universe has found ways to create order out of chaos. Amid chaos, pockets of order appeared surrounded by disorder. All of nature—the entirety of the visible universe—is in an ongoing struggle at every level to escape the gaping abyss of nothingness. According to traditional scientific positivism, the material world consists of inanimate matter composed mostly of inorganic chemistry that, in the words of Terrence Deacon,[2] can be "reasonably well" explained by a "common set of physical equations"[3]. This universe, according to Newtonian physics, is ruled by predestiny and chance of circumstance. The movement of any object simply follows from the external forces of cause and effect placed upon

---

2. Terrence Deacon's, *Incomplete Nature-How Mind Emerged from Matter* 2013 Norton. Kindle Edition.
3. Deacon, *Incomplete Nature* 144.

it. Following the second law, all things are on a march toward extinction. The scientific measure for the grade of disorder is expressed as entropy. Left to its own device, entropy would continuously increase, and the whole universe would eventually end up in thermodynamic equilibrium. Nothing could exist in such a universe—no plants, no animals, and no human beings.

Imagine two adjacent rooms. One room is filled with poisonous gas, the other has an invisible caretaker who has lovingly assembled a luscious green habitat, which is the home of a cute, nearly extinct, cuddly panda. Now imagine some irresponsible keeper leaves the door between those two rooms wide open. The laws of thermodynamics assure us that the poisonous gas would seep into the panda's habitat. We could assume that in desperation the panda would run into the other room, but to no avail. Eventually, our panda, the plants, and everything else alive would wither away. This is what happens when entropy is at its highest level—equilibrium of the gas in both rooms would be achieved, and both rooms would be equally poisonous for every living thing.

If the second law of thermodynamics had its way, the fate of the universe would be worse. It would not leave anything untouched. Everything would disappear—no more cute pandas anywhere, no rocks, no rivers, no forests, and no stars in the night sky. It would be a place without differentiation, the great void, the ultimate nothingness. Throughout the course of the nineteenth century, people began to grasp the implications of the second law, causing a great desperation among scientists. One writer imagined the universe as a giant fish bowl with human beings as the fish. Above it an evil god took pleasure in watching the fish slowly wither away.

Other writers cautioned against accepting this catastrophe as the ultimate fate of the universe. Looking carefully at our immediate world, they concluded that something must slow down the decay and stop, or at least delay, the looming disaster. Living things seemed to counteract the second law. Life has managed to build little islands of order in this disorderly, strange, and dangerous world. How can life be so different?

In his recent book, *Incomplete Nature—How Mind Emerged from Matter*, Terrence Deacon speculates that when living things appeared on earth some three billion years ago, the universe made a "radical about-face." At that point, according to Deacon, a "reversal of typical thermodynamic tendencies" happened that led to a "fundamental reorganization."[4] Deacon claims that at that moment, "the reliable one-dimensional lockstep of just one thing after another, which had exclusively characterized physical events throughout the universe, took an abrupt turn and headed the other way."[5]

The cause for this monumental transition, which evidently counteracted the universal trend toward increasing

---

4. Ibid., 143.
5. Ibid.

disorder, is described by Deacon as an "essentially new form of organization," powerful enough to negate the destructive effects of the second law. This new organizational form, which introduces order, awareness, relationship, and cooperation, is often called negentropy. While modern scientists are by no means in agreement what negentropy really is, or even whether it truly exists, there can be no doubt that with the appearance of the first human beings on this earth, the quest for meaning, for order, the judgment of value and beauty were introduced in a way no one could claim existed here before. Even the most steadfast materialists—those who claim that there can only be matter and that the human mind is the mere firing of electrons and possibly nothing more than an illusion—would have to admit that nonliving material things couldn't even have the illusion of a mind. Therefore, human beings, perhaps other animals, and possibly even plants, are fundamentally different from rocks and stardust. Or are they? Is that really where the evidence leads, or have we made a premature assertion about the nature of reality?

More importantly, what brought this monumental transition about? How could something as complex and mysterious as mind appear out of thin air? How could awareness and consciousness spring out of nothing? How could anything come out of nothing? The answer is collective orchestration. The story I am about to tell in this book is one of relationships, communication, and cooperation.

As we will see, nature has found an escape route. Through effective cooperation, individuals can move as a single unit into a higher level of complexity, leaving uncooperative members behind. It is as if—in the words of social psychologist Jonathan Haidt—"a staircase to a secret door opens to seemingly nowhere."[6] Whenever in the history of the universe a group of individuals effectively cooperates, a door may open to a room no one knew existed. It is still the same house, the same universe. But now there is a new plateau, with new possibilities and new vistas. Those individuals, who have managed to establish themselves on the next higher level, collectively form a new organism, a kind of superorganism that can procreate at the higher level. Through cooperation, what started out as many has become one – in a higher dimension of freedom; e pluribus unum. But they are safe only for a short while. Still living in the same home and obeying the same laws, uncooperative free riders will eventually try to spoil their success. The spoilers keep tearing away at the foundation, while random mutation creates new free riders among the higher rank. Again, escape is possible through renewed cooperation. Individuals work together, and collectively they create yet another new superorganism, a new individual capable of populating the next higher level. And so the game of creation and destruction continues. Eventually, it brings forth our wonderful world, a highly

---

6. "Religion, Evolution, and the Ecstasy of Self-Transcendence," Jonathan Haidt, TED Talks, February 2012, .

complex mansion with many rooms and many floors, and many dimensions. At every level, destruction and death are possible. But those who cooperate can escape, at least for a while. Perhaps everything in the universe is driven by the dim awareness that together, as a group, new vistas are possible and new horizons are within reach. Death can be overcome, if only temporarily. But most of all, the whole universe is in a constant struggle to survive. I hope to make a convincing argument to demonstrate that in the ongoing game of survival, nature favors cooperation over competition, and as a by-product, creates a hierarchy of increasingly more complex systems.

The "new organizational form" Deacon talks about is based on a principle I call collective orchestration. Once aware of it, I found collective orchestration happening at all levels of nature, not only among living things. Gone is Deacon's simplistic model of an early universe that moves in easily understandable regularity. Collective orchestration occurs not only among living things but is part of the evolution of molecules, atoms, and all the other elementary particles. It finally occurred to me that space-time is the result of the successful orchestration of even more elementary space-time particles.

My hypotheses is that in nature, at every level, individuals may cooperate and synchronize, and under certain conditions act as one. A new individual of higher complexity emerges, which fits a phenomenon Darwin called "super-organism." While Darwin used this term only to describe social phenomena that he had observed among insects, I will show that the same process is ubiquitous, producing superorganisms at all levels. A superorganism is a collective that exhibits qualities not present at the level of the lower-grade individuals. These qualities lead to an increase in complexity, which is responsible for the apparent hierarchical order in nature. This process is not a mere accidental result of random mutations or coincidences. Rather, the observed cooperation, synchronization, and creation of super-organisms is nature's strategy to prevent extinction; by effectively counteracting the second law of thermodynamics. Though, never negating it. At every level, there are individuals who have a natural tendency to cooperate. Through random mutation, some individuals may evolve who selfishly take advantage of the cooperating group. Such uncooperative free riders may occur in any group. They can take over and eventually destroy any natural system. Thus, the immense complexity and hierarchical structure in nature is the result of a communal effort—the outcome of a universal cooperative principle I call "collective orchestration."

But why is collective orchestration necessary for evolution? The short answer is: cooperation preserves energy. To avoid premature decay or even death, every organism must preserve energy. The premiere way to preserve energy is through cooperation. By being able to collectively advance, those who cooperate gain a clear advantage in the evolutionary race.

CHAPTER 2

Faust's Wager with the Devil

• • •

"If to the moment I shall ever say..."

THE LEGENDARY CHARACTER FAUST MAKES a wager with the devil, promising to turn over his soul as payment for a moment of bliss. In Johann Wolfgang von Goethe's version of the Faust story, *Mephistopheles*, the prince of darkness, makes every possible attempt to provide Faust with such a moment of eternal bliss. He offers him the innocence of a young love, lets him drink from the fountain of youth, tries to corrupt him with the seduction of imperial power, and finally reveals to him the deepest secrets of nature. The devil even guides him down to the source of all being—the eternal mothers, a place not even the devil is able to visit. Despite this gigantic effort, Faust remains deeply unsatisfied. Nothing could provide him with a moment of complete bliss—a moment that would be so desirable, so perfect, that no other wish remains.

Faust rushes on, from one desire to the next, from one fulfillment to another. "Oh, that in this world naught perfect is, I now do see" he cries out, bemoaning the deep dissatisfaction with God's creation.

Goethe wrote and rewrote his Faust story for most of his adult life. At eighty-two years of age, when he finally passed away, he had completed the second part of his Faust epic. As an aging sage, he found an answer to Faust's quest. In a utopian vision, Faust projects the perfect cooperation of a proud and free people. If such cooperation could ever be achieved among human beings, Faust fantasizes, then he would truly experience the moment of bliss he so desires. At that moment he would gladly turn over his soul to the devil. In isolation, an individual cannot reach bliss. Bliss has to be the result of a communal effort. As a group, we can lift ourselves out of the misery of existence and escape death. Collectively, as a cooperating group of equals, we can move to a higher plain. Not unlike the mystical body of Christ, which is established wherever two or three believers are gathered in His name, Goethe offers a secular, nonsectarian vision of mystical harmony: the harmonious cooperation of working people, who collectively can defeat death and establish perfect harmony.

CHAPTER 3

Wisdom of Crowds—Many Know Better

● ● ●

TODAY MOST OF US ARE embracing an increasingly individualistic culture. As a character, Goethe's Faust is the quintessential individualist. Individualism is a historical construction based on the essential values created during the European Renaissance. Disconnected from tribe and family, Faust is the product of his education, wholly responsible for his actions and in charge of his destiny. Even with all his knowledge and advanced doctoral degrees—Faust studied all the learned subjects of his time—he cannot find satisfaction or happiness. Perhaps, as modern psychology is increasingly demonstrating, happiness cannot be found within the individual, but is somehow deeply connected to relationships and to others.

Western civilization has birthed the individual and has come to deify it. Let me hurry to say that by no means do I advocate deconstruction of the self, nor do I intend to downplay the extraordinary significance of this historical achievement. It is responsible for the incomparable advances that propelled Western civilization to the fore-front of the world's cultures. But we should never forget that individualism and the focus on the autonomous worth of each individual is a rather recent development, unknown to human beings walking the earth for millions of years. Early human beings lived in groups and were identified by their group. Persons derived their worth from the tribe or the family they belonged to. Without that identity, a man's value was his physical strength, or sometimes it was the sheer power of a cunning mind. In the absence of these characteristics, a man was worth nothing. Some of the oldest stories, such as Homer's *Iliad*, were records of battles between tribes, of wars, and epic movements of groups of peoples. Other stories reported the life of heroes such as Odysseus and Gilgamesh, men who distinguished themselves from the rest of the unnamed masses by their physical power and their often-godlike deeds.

Beginning with Homer's writings in the eighth century BCE, classical Greek society began focusing increasingly more on the history, fate, and value of the individual person. Homer's second great epic, the *Odyssey*, narrowly focused on the journey of its hero. Odysseus was known—often in advance—to those he met on his journey, as a man of great knowledge, courage, wisdom, and power. While his traveling companions remained mostly a faceless crowd, cunning Odysseus distinguished himself by his incredible wit, foresight, and by the ability to consume wine without getting drunk. When the witch Circe seduced Odysseus's group with an orgy of alcohol consumption, Odysseus's companions turned into pigs. Odysseus alone remained his unwavering self. Promptly, the witch recognized her famous guest.

By the time the Greek philosophers, the famous founders of Western philosophical discourse (most importantly, Socrates, Plato, and Aristotle), walked the earth, the idea of the individual worth of a person began to take hold in Greek society. From here, the idea of individualism began its triumphant march around the globe. It still attracts new converts and is changing perspectives, especially in Africa and Asia, continents where communal and collective societies have survived the longest. I recently visited the opening of an art show at a gallery in Beijing, China. The painting collection was presented under the theme, *Who am I?* The art show was a study on the growing influence Western ideas of self are having on Chinese society and art. Traditional Chinese art rarely focuses on representations of a single person. More common are depictions of a group of people or nature, in which the human being is depicted as a mere part of the whole.

Under the influence of the great logician Aristotle, Greek philosophy, which initially was conceived as the love of wisdom, turned into what is now most often known as the love of knowledge. This may, on the surface, be only a small semantic shift, but its influence on the way we think and see the world is profound. While wisdom draws its insights from a largely emotional base, which is a substrate of communal experience and generations of practice; knowledge is rooted in firmly established mental constructs. These constructs are the result of individual ruminations, mathematical in character and often believed to contain objective truth. In many of his writings, Aristotle showed a sense and appreciation for ancient wisdom that is deposited in his view of average people. He referred to their insights as the "wisdom of the multitude." But his influence on Western thought came mostly from his ontological foundation of universals and his powerful system of logical deduction.

The chorus in Greek tragedies preserved the memory of ancient crowd wisdom. The chorus often exhibited superior wisdom when compared to the knowledge of its heroes. Judgments in Greek courts were made by a large group of jurors, usually five hundred men, continuing the ancient belief that a group of people could come closer to the truth than individual judges. Today, in the execution of justice, most modern courts trust the rule of law as interpreted by learned judges, while others continue the use of juries, which are made up of ordinary citizens. In drama, neoclassical aesthetics eliminated the role of the chorus and focused narrowly on the treatment of its hero, thus continuing the trend to individualization. In the visual arts, since the Renaissance, the artist zoomed in on the personal features of individual characters, powerful religious or social leaders. In philosophy, Descartes's *cogito ergo sum* decisively directed any quest for certainty and absoluteness to be funneled through the narrow conduit of the individual self. Modernity, the modern world through the eyes of postclassical liberalism, paid homage to the individual, assuming that rationality and reason would not only crack the riddle of existence but also solve the problems found in the foundation of morality itself.

More recently, the belief in the wisdom of the multitude, now call the wisdom of crowds, experienced an interesting comeback. Some psychologists and social scientists began to research knowledge and wisdom that can apparently be found in crowds. James Surowiecki begins his recent book on *The Wisdom of Crowds* by cautioning: "One of the striking things about the wisdom of crowds is that even though its effects are all around us, it's easy to miss, and, even when it's seen, it can be hard to accept."[7]

According to Surowiecki, empirical research shows that consulting a properly composed group of average people may often yield superior results to consulting an expert. Evidently, the key to getting good results lies in the very composition of the group. Groups that are composed of a greater variety of people, such as members of different ethnic, religious, and cultural backgrounds, have a greater chance in achieving credible results than homogeneous groups consisting of members who are quite similar to each other.

---

7. James Surowiecki, *The Wisdom of Crowds* (New York: Random House, 2005).

It would seem that contrary to the popular assumption that individual genius is always superior to the knowledge of average people, the group as a whole might actually know better than any one individual, especially when it is the result of effective cooperation. But what is the situation in the animal world?

When it comes to numbers, it has long been known that insects are the most successful species. Darwin already observed that insects, like no other species in the world, have learned how to cooperate. But are insects also smart? *National Geographic*'s staff writer Peter Miller thinks they are; not as individuals, but as crowds or as colonies. In an article, "The Genius of Swarms,"[8] Miller introduces the astounding abilities and achievements of entire colonies of certain insects such as bees. He credits the tendency to cooperate that gives these animals at least the appearance of a certain group intelligence. Miller asks the provocative question: "Does nature employ cooperation in the same way it uses natural selection, competition, and random mutation?" He says that in search of an answer he "came upon some strange phenomena of synchronized and collective behavior in the world around us."

Could it be that crowd intelligence is not limited to human beings but can readily be found among animals? Collective orchestration seems to unravel the mystery of intelligent crowds.

---

8 Peter Miller, *Swarm Theory*, JULY, 2016

CHAPTER 4

Collective Orchestration—A Personal Journey

● ● ●

MY FIRST INSIGHT INTO THE phenomena of collective behavior and synchronization in nature came from observation. Walking one evening on a circular road in Rocky Mountain National Park, I noticed that crickets in one place were chirping in unison, while a few steps further they sounded dissonant, much like a choir with half the voices out of tune. In one place, it was a cacophony of sounds while a little further away the choir was unified. Curiously, I walked the circle again and observed the same phenomenon. What made the difference between the two groups? To my surprise, half an hour later, walking the circle again, I noticed that the first group suddenly sang in unison as well, and it appeared to me that the volume now was much stronger, more energetic—as if there was joy in their voices for having it finally figured out.

A few months later, I camped overnight on an organic farm. In the morning, at the sight of the first rays of sunlight, an overwhelming orchestra of birds greeted the new day. After about ten minutes, the birds quieted down, and for the rest of the day, they sang separately, calling their mates or simply enjoying the day. Each bird seemed to follow its own path.

This all came back into mind one early morning in the Moroccan city of Marrakech. I was awoken at first break of sunlight over the Atlas Mountains by the voices of muezzins calling to prayer. The subsequent mornings I got up before sunrise and ascended to the rooftop of my hotel, located in the Medina, to observe the singing preachers. Not unlike those birds, they greeted the new morning with their chanting voices. Could it be that human beings have learned from birds when and how to pray? But then I observed something even stranger. There was a morning when one lonely voice started chanting long before sunrise. This muezzin sang for a while, his voice sounding frail and tired. After a few minutes, when nobody joined him in the song, his voice died down and remained silent until the rest of the voices began their sacred chant. On some mornings, the chanting never quite got off the ground and remained unenthusiastic. On other mornings, they all seemed to sing in unison, as if praising the Lord with one voice. The chorus swelled in volume, and the excitement of all participants was palpable. What was going on here? What made the difference? Human beings, just like birds, enjoy sounding off together, and just like those crickets, they become enthused when they finally reach the climax of togetherness. Like an orchestra, they seemed to practice cooperation, and there was satisfaction and joy when they finally found the right tune.

A few years later in Austin, Texas, I observed bats, thousands of them, pouring forth from underneath a bridge. Punctually, every evening at the same time as if an alarm had gone off, they emerged from underneath the bridge in an orderly fashion, like a zipper being unzipped. Ascending into the evening sky, the bats began creating the most amazing patterns and formations. How could hundreds or thousands of them create those beautiful patterns in unison, at exactly the right moment? The whole thing looked like an orchestra conducted by an invisible hand, or a painting in which the animals were the pixels. But this was achieved without an apparent conductor, without a script, without a central command center. Much like a modern jazz ensemble, they created these formations together, collectively.

On a trip to South East Asia, I experienced the same phenomenon again. Only this time it was achieved by thousands, perhaps millions of tiny fireflies, all blinking on and off at the very same moment. I began calling this phenomenon collective orchestration. Once I became hyper aware of this collective orchestration of behavior in the world around me, I started discovering additional examples within the scientific literature. I concluded that in nature, on many occasions, individuals cooperate to form a larger system that tends to act as one. Collective behavior is not just a pattern in nature that human beings happen to find aesthetically pleasing. Collective orchestration is a powerful force of group cohesion that leads to measurable survival benefits for the group. It can be found throughout nature.

CHAPTER 5

For Example, the Dictyostelium

● ● ●

ONE OF THE MOST STRIKING examples for collective orchestration lives in ordinary mud puddles all around us. For millions of years after the first life appeared, only one-cell organisms populated the earth. Then, about 580 million years ago, most of the life forms currently populating our planet developed in a relative short time of just eighty million years. This period is known as the Cambrian explosion, indicating the rather rapid proliferation of life during that time. Darwin was puzzled by the speed with which the current diversity of life had developed. He believed this to be a serious argument against his theory of evolution, which was based on mutation and natural selection. According to his theory, life developed incrementally, not in giant leaps. What had happened here? How did life evolve from a one-cell organism to the great variety of multicellular organisms we know today? The answer may be collective orchestration. To find examples for this miraculous process, we don't have to look very far. It happens daily in the world around us. Simple one-cell creatures band together and act as one larger organism, a superorganism if you will.

The life cycle of this tiny organism called the "dictyostelium" is peculiar. Normally living in mud puddles, this one-cell organism exists in isolation for most of its short life cycle. As much as scientists can tell, each of these tiny creatures is identical to the other. A strange phenomenon occurs when a qualified number of them experiences a deficiency in food and living space. Scientists have observed how they suddenly group together to cooperate in a formerly unknown way. Collectively, they form one large, multicellular organism with a head, a tail, and a primitive digestive system. Each cell takes over a specific function within the new organism. They evidently group together to achieve goals that they could not achieve by themselves. The new body enables this collection of individual cells to move to new feeding grounds. They accomplish this in several metamorphic stages.

After several hours, the dictyostelium slug goes through another change. The back end catches up with the tip, and the slug turns into a blob. The blob stretches upward a second time, and now some amoeba produce rigid bundles of cellulose. They die in the process, but their sacrifice allows the blob to become a slender stalk. Perched atop the stalk is a globe, bulging with living amoebas, each of which covers itself in a cellulose coat and becomes a dormant spore. In this form, the colony will wait until something—a drop of rainwater, a passing worm, the foot of a bird—picks up the spores and takes them to a bacteria-rich place where they can emerge from their shells and start their lives over.

Today scientists do not know what makes some cells become tails, others heads, and yet others digestive organs. There is no observable difference in the original units. As far as scientists have observed, they are identical. Yet, at the moment of unification they "know" and perhaps even "choose" their places. Scientists have speculated much about this phenomenon.

Collective orchestration offers a somewhat different interpretation of the same event. Running out of food (that is, out of energy) those amoeba do what other organisms would do under similar conditions—they huddle together. We do not have to project great planning for this to work. All these spores have to do is cluster together and wait until the little stalk is carried away into a different puddle with new feeding grounds. These patient dictyostelium will succeed in the evolutionary race. The initial cause for the whole process to begin is the lack of energy.

Other scientists have observed this seemingly social behavior of single-celled bacteria and called it "cooperative dispersal." These scientists found evidence of other forms of complex behavior at the level of microorganism as well. Bernard Cresp, from the Department of

Biosciences at Simon Fraser University, in an article titled "The Evolution of Social Behavior in Microorganisms,"[9] writes:

> Recent studies of microorganisms have revealed diverse complex social behaviors, including cooperation in foraging, building, reproducing, dispersing, and communicating. These microorganisms should provide novel, tractable systems for the analysis of social evolution.[10]

In the face of an apparent food shortage, some of these bacteria have been observed to undergo "programmed death" or suicide in an attempt to provide food for the remaining group.[11] What is the source of such "social" behavior—a label, which until recently, would have been called undue anthropomorphizing? I believe that traditional modes and explanations are unable to account for such phenomena. And yet they have clearly been observed and documented. Collective orchestration provides a reasonable explanation.

---

9 The Evolution of Social Behavior in Microorganisms," TRENDS in Ecology & *Evolution* Vol.16 No.4 April 2001
10. Ibid., 178.
11. Ibid., 179.

CHAPTER 6

The Emergence of Life—A Communal Effort

● ● ●

THE FAMOUS PHYSICIST ERWIN SCHRÖDINGER was on a quest. He wanted to know how life began. To this effect, more than half a century ago, Schrödinger authored a little book with the promising title: *What is Life?*[12] As a specialist in physics, he was aware that the academic world of his time would offer him little credence to write about a subject that was quite obviously an intricate part of another academic field, biology. But Schrödinger sensed that the mystery of life and its origins on earth could not be solved focusing on one discipline alone. Interdisciplinary research on the other hand, was not very common in his time. The rule for academics was quite frankly to stay within their field. But for Schrödinger, it was clear that a certain interdisciplinary approach was needed to solve the mystery of the origin of life. What began one day as part of chemistry, the next day would

---

12. Erwin Schrödinger, *What is Life?* (Cambridge: Cambridge University Press, 1944).

become part of biology. It was that simple. Life had to come from chemistry. For a dedicated scientist, there was no other solution. Life must have a material base. Nothing else could be considered science.

Needless to say, Schrödinger's search remained unsuccessful. As a scientist of his time, he was committed to the reductionist method. As a reductionist, he believed that knowing the smallest particles of any complex structure would yield the secrets of that structure. Nothing qualitatively was added when simple things became more complex. The whole was always considered to be the sum of its parts. Schrödinger believed that his detailed research into the world of atoms and nuclear particles would ultimately show him the secrets of life. In hindsight, one could say that Schrödinger was on the right track. Contrary to the commonly held beliefs that life was the result of divine intervention, the emergence of life had to be part of evolution at the material base. There must be a seamless development from the smallest particles to the complexity of the human brain. The steps are incremental, much more so than the academic division of the world into living organic matter and nonliving, inorganic matter would have you believe. This division between organic and inorganic goes back to one of the gravest mistakes the ancient Greeks made. They believed that everything that moved by itself had a soul and was therefore part of biology. All inert things, those that did not move by themselves, were part of the material world, which the ancient philosophers called inanimate, without a soul. Only the animate world was created by the gods. Inanimate nature has either always been there or was made by lower gods. Christianity adopted this arbitrary division. Today it continues in the various disciplines of the academic world.

Schrödinger was correct when he crossed disciplines to investigate the origins of life. What he could not know was that the very method he used to analyze nature was holding him back from developing successful conclusions. Not only was the nature of DNA unknown in his time, scientists had not yet realized the enormous complexity that emerges when natural systems cooperate. Nor how they produce unpredictable new qualities that forever define the newly developed, higher order system. Complex biological life is one of those emergent properties that could not—and never will—be understood by analyzing its constituent parts alone. Like the dissecting of the chemical molecules from which it is made.

A new science, called systems theory, had to be developed that focused on the analysis and the properties of systems. Derived from observations made on artificial constructs, this theory found that as new systems emerge from old ones, they often show qualities that were simply nonexistent at the level of the constituent parts. Just as it is impossible from the tuning of an orchestra to predict which symphony the orchestra will eventually play, so too is it impossible to observe the lower-level parts of a naturally emerging system and accurately predict which new entity of nature is about to evolve. Reductionists could conceivably claim that we simply do not yet know enough about the qualities of the individual parts, and if we had all the information, we could predict the outcome, since in their understanding the whole is never more than the sum of its parts. System theorists, however, can convincingly demonstrate that regardless of how much we know about the individual particles, this knowledge can rarely explain the qualities of the new whole. Through the process of cooperation between millions and trillions of particles, something completely new emerges— unknowable and unpredictable from the vantage point of the lower perspective. Just like entropy effectively destroys order, cooperative systems create order; an order with new qualities that are irreducible to its constituents.

Fifty years after Schrödinger's publication, a new group of scientists—among them physicists, biologists, and system theorists—collaborated to ask the same question.[13] In his contribution *What is Life? Fifty Years Later, Was Schrödinger Right?*[14] the eminent biologist Stuart Kauffman emphasized the importance of collective dynamics in the emergence of life.

Kauffman boldly states: "remarkably, not random mutation or the evolution of DNA, but collective and

---

13. Michael P. Murphy, ed., *What Is Life? The Next Fifty Years* (Otago University Press, New Zealand, 1995).
14. Stuart Kauffman, *What Is Life? Fifty Years Later, Was Schrödinger Right?* 1995.

cooperative, self-organizing processes at the molecular level are now the most likely candidates for the beginning of life." Elsewhere he says: "the ultimate source of order and self-reproduction may lie in the emergence of collectively ordered dynamics in complex chemical reaction systems." Kauffman concludes: "Such complex reaction systems can spontaneously cross a threshold, or phase transition, beyond which they become capable of collective self-reproduction, evolution, and exquisitely ordered dynamical behavior. The ultimate sources of the order requisite for life's emergence and evolution may rest on new principles of collective emergent behavior in far from equilibrium reaction systems."[15]

It appears now that life may not be the result of one molecule mutating into a new quality called life and then successfully replicating, but that life is the result of a collaborative effort of millions of particles, or elements. If life is the result of collective behavior, then perhaps collective behavior is more important and more essential to the whole evolutionary process than previously assumed.

Biophysicist Jeremy England of MIT agrees: "You start with a random clump of atoms, and if you shine light on it for long enough, it should not be so surprising that you get a plant."[16] According to England, "there is one

---

15. Kauffman, *What Is Life? Fifty Years Later, Was Schrödinger Right?* 84.

16 Natalie Wolchover, "A New Physics Theory of Life," Quanta, January 22, 2014.

essential difference between living things and inanimate clumps of carbon atoms: The former tend to be much better at capturing energy from their environment and dissipating that energy as heat." England offers a mathematical formula that "indicates that when a group of atoms is driven by an external source of energy (like the sun or chemical fuel) and surrounded by a heat bath (like the ocean or atmosphere), it will often gradually restructure itself in order to dissipate increasingly more energy. This could mean that under certain conditions, matter inexorably acquires the key physical attribute associated with life." A theory he calls "Dissipation-Driven Adaptation".

Collective orchestration offers an addendum to this explanation: A random clump of atoms experiences a significant change in the availability of energy. The individual atoms begin to relate to each other in the form of synchronized vibrations. Through effective cooperation, the newly complex system saves energy. The collective symphony of atoms generates a surplus of available energy, which is utilized by the newly created superorganism and dissipated as heat. England is right: add energy to it, and under the right conditions, life will evolve.

Those "right conditions" will most certainly include cooperation of the constituent particles. Cooperation often requires synchronization. Let's investigate the difference between synchronization and cooperation.

CHAPTER 7

Synchronization and Cooperation

● ● ●

COOPERATIVE BEHAVIOR OFTEN BEGINS WITH a group of individuals synchronizing their actions. In *Sync: The Emerging Science of Spontaneous Order*, the mathematician Steven Strogatz[17] took a detailed inventory of synchronized behavior at all levels of nature. Strogatz found hundreds of examples for synchronized behavior among animals. Humans, birds, fish, and fireflies; all following the same synchronized patterns. Additionally, he found these patterns in tiny cells, such as the pacemaker cells that propel the human heart and neurons in our brain. Other cells synchronize to build skin or the pancreas. Says Strogatz: "Groups of fireflies, planets, or pacemaker cells are all collections of oscillators— entities that cycle automatically, that repeat themselves over and over again at more or less regular time intervals...nature uses every available channel to allow these

---

17 Steven Strogatz, *Sync: The Emerging Science of Spontaneous Order*, 2003.

oscillators to talk to one another...and the result of those conversations is often synchrony, in which all the oscillators begin to move as one."[18]

As it appears many animals as well as human beings sometimes synchronize their actions. But why? What purpose does it serve? And does this also mean they cooperate? This is a much more difficult question. The examples I have given so far to illustrate collective orchestration included synchronization as well as the type of cooperation that leads to the formation of a group or crowd behaving as if they were one. In order to get a clearer understanding of what is going on here, we need to clarify the terms cooperation and synchronization. Often, even in academic writings, these are used as if both meant the same thing. Cooperation is generally defined as "an act or instance of working or acting together for a common purpose or benefit."[19] We have a clear understanding when we say: "people cooperate". When human beings cooperate, we assume some element of choice, perhaps even altruism; the selfless care for others. But blinking fireflies or swarming birds? These may be synchronized, but we don't necessarily assume that these animals also cooperate. Synchronization is defined as "to occur at the same

---

18 Steven H. Strogatz, *Sync: How Order Emerges from Chaos In the Universe, Nature, and Daily Life* (Kindle Location 74). Kindle Edition. (Hachette Books, 2012).

19 http://dictionary.reference.com/browse/cooperation, accessed April 12, 2013.

time."[20] Under certain conditions individuals synchronize their actions to achieve a collective goal. But in many instances, especially among so-called lifeless things, the reason for synchronization is not clear.

Among human beings and animals, we find synchronization as well as cooperation. What comes first—cooperation or synchronization? Or do both indicate the same phenomenon? How is cooperation related to synchronization? Under what circumstances does synchronization turn into cooperation? For Strogatz, the question does not occur because he uses both terms almost interchangeably.

While synchronization seems to involve a somewhat mechanical process, cooperation often appears to be goal-oriented, the result of deliberate action. Synchronization is possible without cooperation. While effective cooperation may require synchronization, synchronization does not always require cooperation. If all clocks in a building show the same time, the clocks are synchronized, but they do not cooperate. We can also imagine cooperation without synchronization. A basketball team will have to cooperate to win a game, but they don't necessarily have to synchronize their moves. Cheerleaders, on the other hand, would look silly if their movements were not carefully synchronized. But both teams will, perhaps in a less obvious sense, also cooperate. When we see synchronized action, do we immediately assume cooperation? No.

---

20 http://www.thefreedictionary.com/synchronizationm, accessed April 10, 2013.

If a synchronized action is the result of a choice made by each member of a group we tend to refer to this as cooperation. But if synchronization is coerced, if the members are forced to synchronize, there likely is no cooperation. We also do not speak of cooperation when synchronization is the result of a mechanical process; therefore, if synchronization is to appear as cooperation, the synchronized action must be somehow the result of a choice. To be capable of making a choice, there must be some agency present in each synchronizing individual. Within the group of cheerleaders, all individuals would have made a decision to cooperate and to synchronize their movements with the other cheerleaders. Synchronization often includes an aesthetic dimension. We admire and enjoy synchronized actions. Just think of the synchronized flying of fighter planes in an air show. We know that the planes fly in perfect formation because skilled pilots direct them. Looking up in the sky, we often see large birds fly in a perfect formation. Who directs them? Furthermore, who directs the synchronized actions of millions of heart pacemaker cells? Is this the result of cooperation or of some mechanical process? Today, an artificial pacemaker can replace the regular rhythm of the human heart when the cells themselves fail to provide it. Is the regular beat of the human heart the result of cooperation? Even more mysterious is the cooperation of millions of photons synchronizing to produce a laser light. When Strogatz observed this, he wrote:

It's surprising enough to see animals cooperating—thousands of crickets chirping in unison on a summer night; the graceful undulating of schools of fish—but it's even more shocking to see mobs of mindless things falling into step by themselves. These phenomena are so incredible that some commentators have been led to deny their existence, attributing them to illusions, accidents, or perceptual errors. Other observers have soared into mysticism, attributing sync to supernatural forces in the cosmos.[21]

Why are synchronized actions of inanimate objects so unbelievable that observers deny their existence or, worse, fell back on magical explanations? This is because when random objects begin to synchronize, this action, at least on the surface, seems to be purposeful. Some agent with intention seems to be present. Yet, these synchronizations don't appear to be directed. If we do not recognize a cause for an action or phenomenon, ordinary people often invoke supernatural forces; whereas scientists prefer to deny the existence of the phenomenon. Both responses are, of course, unsatisfying. Furthermore, when animals synchronize Strogatz calls it "cooperation," but when inanimate objects synchronize, he refers to it as "falling into step" as if the synchronization was the result

---

21 Strogatz, *Sync: How Order Emerges From Chaos in the Universe, Nature, and Daily Life*, 2–3.

of a mechanical, random occurrence. On the other hand, Strogatz also talks about particles "communicating" with each other. If we assert that cooperation requires an agency, then so too, must communication. What's going on? What causes this confusion?

This question becomes specifically relevant with regard to the beginning of life. If we accept the premise advocated in Jeremy England's work; that a random clump of atoms will turn into a living entity under certain conditions; then we must investigate at what moment mere synchronization turns into planned cooperation. Traditional science does not ascribe life to a clump of atoms. So, either those atoms cooperate to synchronize or they synchronize accidentally and then begin to cooperate. A clump of atoms could be called an aggregate. These atoms do not constitute a system. However, when life occurs, we start talking about a system.

To shed more light on this situation, we will have to make a clear distinction between systems and aggregates.

CHAPTER 8

Systems and Aggregates

● ● ●

WHEN LUDWIG VON BERTALANFFY FIRST conceived of a general system theory, he kept the definition of what actually makes a system purposely vague. His expressed goal was to find common laws or rules that were true for all systems, whether created by human beings or by nature. System theory seeks to discover whether all the diverse things we describe as systems share common characteristics? In this broad sense, systems are found in almost every discipline. This includes organizational systems, such as those in biology, ecology, and many other fields. In this general sense, a system may be defined as composed of regularly interacting or interrelating parts.

For our purpose, we define a system more narrowly as a living, interacting, and communicating unit that comprises a number of subsystems, which again are compositions of yet smaller systems. The larger system—for example, the human body—is the result of successful integration of a myriad of smaller systems. The emphasis here is on successful. There are many systems, which are still in the process of further integration. Only few succeed and reach a level of integration that brings about properties only achieved at the group level. Those who apparently fail are aggregates. These are lower-level systems that are in the process of integration, but are more or less far away from reaching the goal of higher-order integration. There are of course many collections of items, which never will be, or aspire to be, a functioning system. So, a collection of stamps is not a system, even though the boy who collects the stamps may have an intricate system of ordering them. Likewise, the pages of *War and Peace* are not a system. All these things are aggregates, not a system; at least, according to our definition. Brian Greene asks the provocative question: What happens if one throws the 670 pages of *War and Peace* in the air. Will they ever return in the same order again? The probability of this happening is infinitesimally small, though not impossible. This statistically nearly impossible proposition gave rise to wild speculations of alternate universes, multiverses, and other unlikely monstrosities.

According to crowd sourced knowledge of Wikipedia, an aggregate is "a collection of items that are gathered together to form a total quantity." The emphasis here is on quantity. The collected items have not yet or perhaps never will reach the transition point, also called phase transition, at which new qualities appear that were previously unimaginable at the former level. The cooperating subsystems collectively reach the level at which one could speak of the creation of a new individual, a superorganism, a system of a higher order. It is the point at which the cacophony of a tuning orchestra breaks into a beautiful symphony, the point at which all the individual parts suddenly begin to act as one. At this juncture, the newly born individual is capable of performing tasks that were unthinkable, unimaginable at the level of the lower individuals.

The universe thus is composed of systems and aggregates. That's it. There is nothing else. So, for instance, a rock is an aggregate of systems that are bound together at the molecular level. The rock's individual parts do not communicate with each other. If one part of a rock gets chipped off, the rest of the rock takes no notice, except at the exact surface where the molecular bonds of the break occurred. At the level of molecules there is communication, communication in the form of chemical bonds. More importantly, each molecule is its own little universe. Each part communicates with the other. A rock therefore could be said to be an unsuccessful attempt in molecular communication. Now take a tree in contrast. We can safely assume that each part of a living tree communicates with all other parts. When the roots lack water, the leaves start to wilt. As a living plant, a tree is an example for a system-wide molecular-communication network and thus a successful cooperation of molecules. But a molecule itself is the result of successful cooperation of yet smaller systems comprising atoms. Each molecule is the result of the combination of two or more atoms. Each atom again is its own little universe in which each part communicates and is connected to the others.

An atom is invisible to the naked eye; so how do we know about the inner life of an atom? Until quite recently the atom was a theoretical particle, assumed to be the smallest indivisible part of the material world. The Greek atomists believed that *materia* could be divided into ever smaller parts, but at one point it would be impossible to divide any further because the tool to cut would not be sharp enough. This part they called the atom, the indivisible. This remained true until the beginning of the twentieth century when the inner life of an atom was discovered. This lead eventually to the invention of the atom bomb. Splitting the atom, the smallest system known at the time, was shown to release a tremendous amount of energy, especially when a chain reaction was created.

Have we arrived at the smallest system? No. Each atom consists of smaller particles, such as electrons and neutrons. These are called fundamental particles. But as was soon discovered, these so-called fundamental particles were compounds of yet smaller systems. This quite seriously challenges our ability to measure and observe because the instruments we measure with are made of the same stuff. At the smallest observable level, systems seem to be nested in yet smaller systems. At the smallest level, as on the large scale, everything seems to be in flux, building systems within systems, in a process we call self-organization.

As we will see, a system is fundamentally defined by its ability to self-organize. Through the process of self-organization, disjoined individuals can join efforts and tackle tasks they were unable to do before. Collective orchestration tells us that through the process of self-organization, individuals can become one and collectively move to a higher more complex level of existence.

CHAPTER 9

Self-Organization—Nature's Secret Strategy

● ● ●

THE HUMAN BODY IS AN excellent example of a self-organizing system. In fact, it organizes its countless subsystems until, at the peak of this arrangement, it establishes the most complex thing we know—the self. This is the power of self-organization.

Every morning when we awake, our body reorganizes, reorchestrates, this awareness we call self. This is the first indication that the self is not something static; it rather is in constant flux, always developing, always evolving. Looking at the different stages of a growing child, we can observe that along with the physical growth, the self is also growing. The self is a carefully nourished social construct. Though it is in continuous dialogue with its environment, it is also fully self-organized. It is not directed or controlled by any agent, either from the inside or the outside. Millions of neurons cooperate, building intricate networks that collectively establish the human self, again and again, each time a little different, yet also the same.

Self-organization is defined as "a process where some form of global order or coordination arises out of the local interactions between the components of an initially disordered system."[22] This process is spontaneous, which means that "it is not directed or controlled by any agent or subsystem inside or outside of the system."

In the mythologies of early animistic societies, we can find many levels of implied self-organization. The Egyptian treatment of chaos as a self-creating force comes to mind. While the Greek atomists saw the material world as chaotic, they were the first to develop a theoretical explanation for the transition from chaos to order. They believed that eventually some kind of organization would develop, provided there was enough time and space. They credited the self-organizing ability of the material world to achieve this task. Under the authority of Plato and Aristotle, this changed dramatically. An idealistic model replaced the self-creating materialism of the atomists. Later enforced by Catholic church doctrine, the standard way of thinking was that nature is designed and created by God and remained more or less under God's continuous guidance. Since God was sufficient reason enough to justify existence, self-organization was not needed.

Rene Descartes seems an unlikely candidate to have been one of the first philosophers to seriously propose the concept of self-organization. It fits, however, within a rationalist, deistic frame of mind to assume that *materia*

22 Wikipedia. *https://en.wikipedia.org/wiki/Self-organization*

gradually evolves into ever-higher forms of complexity without specific and continuous guidance, albeit according to a set of fixed laws. In his famous *Discourse on Method*, Descartes argued that God had created the world, not as it is today, but as chaos. This chaos, along with a well-established set of laws that, according to Descartes, not even God was able to alter, evolved according "to its wont." Descartes continues:

> We might well believe, without doing outrage to the miracle of creation, that by this means alone all things which are purely material might in course of time have become such as we observe them to be at present; and their nature is much easier to understand when we see them coming to pass little by little in this manner, as were we to consider them as all complete to begin with.[23]

Under the influence of the scientific theory, beginning with the deists, the belief continued that universal laws inherent in the physical world govern its development. Most of the universe followed those laws in blind obedience. Freedom, the ability to make choices, was only conceded to the human mind, to which it was specifically

---

23 Rene Descartes, "Discourse on the Method of Rightly Conducting the Reason and Seeking for Truth in the Sciences," in *Great Books of the Western World*, vol. 31 (Chicago; The University of Chicago Press, 1952) 55–56.

granted by God. The rest of the world had to follow those ironclad laws of nature. By the end of the nineteenth century, scientists saw the universe as a mechanical construction with human beings holding the only exception. In such a mechanical universe there was no free will and no spontaneity.

With the discovery of quantum mechanics, a new light was shed on the idea of spontaneity. Some of the most important processes in quantum mechanics appeared to occur spontaneously. This put quantum mechanics truly at odds with the idea of a mechanical universe. How could a mechanical universe be built on a sea of uncertainty? The smallest particles that indeed constitute our real world at times didn't appear to be real at all. They seemed to appear and disappear at will, and their position and velocity could never be determined at the same time.

During the middle of the twentieth century, self-organization became the focal point of a new science called cybernetics. Self-organization was first used as a technical term to describe the behavior of automatic systems in artificial intelligence. From here it made its way into a large variety of fields. Applied to the life sciences, it held the promise of making these "soft" academic disciplines more scientific.

It was not until the late seventies and eighties of the twentieth century that the idea of self-organization became more widely used, as scientists began to apply it in describing the self-generating and autocatalytic behavior of complex systems in nature—both living and nonliving. Its popularization is mostly due to the rapid spread of another new science, the theory of chaotic or nonlinear systems; in short, chaos theory.

Chaos theory demonstrated how a degree of orderliness, even beauty, may under certain conditions develop seemingly "out of nowhere." In a world largely without God, the idea of Descartes and the Atomists may not have been so far off the mark. The spontaneous development of a new order seemed to be no longer farfetched.

According to the rules of classical mechanics that Newton developed, physical actions must have a physical cause. Classical physics states that an object in motion will maintain this movement until an outside force interferes and changes the trajectory of the object. Such actions can be described with linear mathematics. A small cause has a small effect, and a big cause correspondingly has a big effect. In the classical world, when describing complex systems, small causes could often be completely ignored. When Newton calculated the movement of the moon around the earth, he ignored the influence of the myriad tugs that millions of other objects in the universe continuously exert on the movement of earth and moon. Since they were rather small, Newton could disregard them. Still his linear calculations yielded an adequate result.

Once chaos theory was fully understood, it became clear to scientists that in systems at the edge of chaos, a completely different set of rules had to be applied. Such systems follow nonlinear mathematics. According to the rules of nonlinearity, a small initial cause can feed back on itself and in a relatively short time dominate the system. A nonlinear system, therefore, is infinitely more dependent on initial conditions than on a linear system. This is expressed in the widely known butterfly effect: "When a butterfly flaps its wings in Hong Kong it can cause a storm in Mexico."

As it turned out, living and self-organizing systems in general follow such a nonlinear logic. According to Strogatz "all problems about self-organization are fundamentally nonlinear. The study of sync has always been entwined with the study of nonlinearity. This synergistic character of nonlinear systems is also what makes them so rich. Every major unsolved problem in science, from consciousness to cancer to the collective craziness of the economy, is nonlinear."[24]

How can it be that human beings who can obviously act spontaneously are separated by a wall of mechanical necessity from the smallest particles of this world, which evidently also act spontaneously? As it turns out, self-organizing systems can be found throughout the whole spectrum of the living and even the nonliving world. This brings spontaneity and spontaneous creation to the center of attention.

---

24 Strogatz, *Sync: How Order Emerges from Chaos in the Universe, Nature, and Daily Life*, 182.

Reflecting on self-organization, Strogatz speaks about "nature's eerie yearning for order." He says: "In every case, these feats of synchrony occur spontaneously, almost as if nature has an eerie yearning for order. And that raises a profound mystery. Scientists have been long baffled by the existence of spontaneous order in the universe. The laws of thermodynamics seem to dictate the opposite,"[25] and elsewhere he says: "The richness of the world around us is due, in large part, to the miracle of self-organization."

As the biologist Kauffman observed, the very idea of self-organization makes us rethink our understanding of a living creature. He says:

> Organisms, we have come to believe, are tinkered together contraptions, ad hoc marriages of design principles, chance, and necessity. I think this view is inadequate. Darwin did not know the power of self-organization. Indeed, we hardly glimpse that power ourselves. Such self-organization, from the origin of life to its coherent dynamics, must play an essential role in this history of life, indeed, I would argue, in any history of life. But Darwin was also correct. Natural selection is always acting. Thus, we must rethink evolutionary theory. The natural history of life is some form of marriage between

---

25 Ibid., 1.

self-organization and selection. We must see life anew and fathom new laws for its unfolding.[26]

Perhaps it is not enough to rethink just the natural history of life but of the whole universe. Collective orchestration suggests that limited aspects of natural selection are not only acting among living things but among everything. When a system is put in the role of fighting for survival, this process can be likened to natural selection. It determines which systems get to replicate and conserve energy for their constituents and which systems fail. Self-organization plays the role of replication by the individuals as they move from system to system. The principles of self-organization may be guiding a version of natural selection that can be applied not only to living things, but to all things.

---

26 Stuart Kauffman, *At Home in the Universe, the Search for Laws of Self-Organization and Complexity*, 1995.

CHAPTER 10

Cooperation in the Inorganic World

● ● ●

How CAN DEAD THINGS COLLABORATE and form groups? When seemingly lifeless things were observed synchronizing and self-organizing, scientists were truly baffled. When Strogatz observed such inanimate objects spontaneously synchronize, he was stunned. We may understand when we find neurons and pacemaker cells synchronize and cooperate; they are part of a living system, which makes this somewhat more plausible. But it gets totally counterintuitive when it comes to only loosely connected molecules, atoms, or photons. But Strogatz found plenty of synchronized behavior among these so-called inanimate objects, and among supposedly dead things. Electrons, photons, molecules of all kinds, under certain conditions, synchronize, self-organize, and cooperate. In the process, they often create a new order. In many cases, at the height of cooperation, they seem to act as one.

Strogatz observed: "It's surprising enough to see animals cooperating—thousands of crickets chirping in unison on a summer night; the graceful undulating of schools of fish, but it's even more shocking to see mobs of mindless things falling into step by themselves."[27] According to Strogatz, when sync occurs among unconscious entities like electrons or cells, "it seems almost miraculous." These phenomena, Strogatz contends "are so incredible that commentators have been led to deny their existence, attributing them to illusions, accidents, or perceptual errors. Other observers have soared into mysticism, attributing sync to supernatural forces in the cosmos."[28]

This scientific bewilderment comes from the perceived lack of agency when we speak about lifeless things. It was relatively easy to imagine self-organization among living things, as there was always an agency present that could initiate and carry out the newly developing order. But how can so-called dead things cooperate and self-organize if there is no agent present to carry out this action?

As a mathematician and conventional scientist, Strogatz found a plausible technological, mechanical explanation for this process of synchronization. According to Strogatz, all these various systems, dead or alive, follow the same mechanism, which is defined by an oscillator, a pulsating device mostly used for the purpose of generating a signal. "Coupled oscillators" says Strogatz, "are systems of such

---

27 Strogatz
28 Ibid.

devices with two or more members that are communicating with each other. Often their communication results in synchronized behavior."[29]

Strogatz seems to have little interest in asking why particles would synchronize in such a way and why these millions of individual photons suddenly "move as one." He seems content in finding a mechanical explanation for the process. Dryly he concludes: "For reasons we don't yet understand, the tendency to synchronize is one of the most pervasive drives in the universe, extending from atoms to animals, from people to planets."[30]

Strogatz's explanation crediting cooperation in the inorganic world to the doings of coupled oscillators is utterly unconvincing. An engineered device can hardly explain an organic process, especially when he also claims that these particles communicate with each other.

Scott Kelso and Hermann Haken, in "Synergetics of Brain and Behavior," give an example of spontaneous change in the micro world. They observe that in dynamical systems "order appears seemingly out of nowhere." It is noteworthy that the change from randomness to organization appears not gradually but abruptly. When a certain threshold is reached the microscopically small particles synchronize or cooperate and make the pattern these particles create macroscopically visible. In the words of the authors: "an amazing event called instability occurs.

29 Ibid., pg. 3.
30 Ibid., 14.

The liquid begins to move microscopically in an orderly rolling motion. The system is no longer a haphazard collection of randomly moving molecules: billions of molecules cooperate to create macroscopic patterns evolving in space and time."[31]

In "At Home in the Universe," Kauffman concludes that "the search for laws of complexity that govern how life arose naturally from a soup of molecules, evolving into the biosphere we see today. Whether we are talking about molecules cooperating to form cells or organisms cooperating to form ecosystems…we will find grounds to believe that Darwinism is not enough, that natural selection cannot be the sole source of the order we see in the world. In crafting the living world, selection has always acted on systems that exhibit spontaneous order. If I am right, this underlying order, further honed by selection, augurs a new place for us—expected, rather than vastly improbable, at home in the universe in a newly understood way."[32]

Could it be that evidence of self-organization brings something much deeper to light? Could it be that the presence of self-organization reveals the existence of operating systems, in which change does not come from the outside ("contragrade" in Deacon's terminology) but change is orthograde—coming from the inside of the system.

---

31 J. A. Scott Kelso and Hermann Haken, "Synergetics of Brain and Behavior," in *What is Life? The Next Fifty Years.* 137–60.
32 Stuart Kauffman, *At Home in the Universe*, Preface.

The current scientific paradigm tends to grant no agency or feelings to most of the universe. This obvious failing may be the deepest reason why a growing number of scientists are re-exploring the ancient philosophy of panpsychism.

CHAPTER 11

The Role of Panpsychism

• • •

THE PHILOSOPHER AND POET GOETHE believed that not only was the biological world organic but so was the whole universe. He and many poets of the romantic tradition saw a fluidity of form permeating all things. The glue that those poets saw connecting form and content was a deeply felt belief in the spirituality of everything. Advancements in the science of physics science, caused all these pantheistic remnants of an earlier time to be discredited and eradicated. Gone forever was the platonic belief in the superiority of form over content. With the discovery of quantum mechanics, and fifty years later, the new emphasis on self-organizing systems, the difference between organic and inorganic seems to be blurred. As Strogatz found out, synchronization and self-organization are not limited to the organic world. These can be found in the inorganic world down to the level of atoms and molecules. But from the materialist perspective, synchronization and self-organization among otherwise nonliving entities appeared to be weird if not somewhat miraculous. A new philosophical approach is needed. The ancient philosophy of panpsychism may be the answer.

For a workshop I conducted on panpsychism at the Center for Consciousness Study at the University of Arizona in Tucson, the organizers had placed a rock in the middle of the stage. Participants were asked whether they believed that this rock was conscious.

According to David Skrbina, panpsychism "is the view that all things have a mind, or a mind-like quality."[33] In other words, contrary to the current materialist view, panpsychists believe that everything that exists is more or less conscious.[34] All things are sentient. Wherever there is matter there is some kind of mental agency present. The universe is not an unconscious machine where mind is a rare and accidental occurrence, but the whole universe is *materia* becoming conscious. Adopting this position has become increasingly more popular among consciousness researchers. This is for good reasons. Panpsychism seems in one swoop to solve some of the most difficult problems plaguing modern consciousness research.

Accepting panpsychism would specifically solve the so-called hard problem, which David Chalmers describes

---

33 David Skrbina, "Panpsychism as an Underlying Theme in Western Philosophy" in *Journal for Consciousness Studies*, 10, no. 3 (2003): 4.

34 I prefer to use the term sentient because consciousness often implies a complex form of mind.

as the answer to the question: What is a conscious experience? Experiencing is what systems do. Aggregates in contrast do not have a uniting experience, neither of themselves nor of their surroundings. So, what is the answer to the above question, whether panpsychists should believe a rock thinks? A rock is not a coherent system, but an aggregate, a connected but not communicating "clump of atoms."

Panpsychism is an ancient worldview common to all indigenous civilization. To this day the majority of all people in the world have some kind of panpsychist worldview. The term itself goes back to two Greek words: *panta* and *psychae* meaning "all" and "soul" or "spirit."

The very fact that panpsychism is an ancient worldview proves somewhat problematic. As Skrbina pointed out, panpsychism has a long history and means slightly different things to different people. A discussion about panpsychism often ends in a debate over what it really means. Some actually believe that a rock is conscious; others believe the whole universe is one conscious mind. In the search for consciousness all panpsychists share the view that *materia* and consciousness are never really separate. But if there are many minds, then the question is: What sets the boundaries? If there is only one mind we must ask why each of us feels the need to be a separate entity. Why don't we all feel as one? Panpsychism claims that everything is sentient and that there are either many separate minds or one single mind that unites everything that is.

In his recent book on panpsychism, Skrbina pointed out that a large group of Western philosophers held the panpsychist view, or at least subscribed to some variety of it.[35] When it comes to poets, religious people, or just plain ordinary folk, panpsychism is almost universally accepted. This holds specifically true for non-Western cultures. Virtually all philosophical speculations and cultural ideologies from non-Western civilizations such as China, India, and West African Vodun, hold predominantly panpsychist views. This would easily make panpsychism the majority view of the world's population. Despite this popularity, most Western scientists seem to despise the panpsychist worldview. What is at the root of such a disconnect? What happened in Western thought that makes the panpsychist view so unacceptable, and why is panpsychism a virtual stranger to the current scientific paradigm?

Western philosophy by and large followed the Greek classical separation of mind and body. In Christianity, this conviction merged with the Hebrew mythological or religious assumption that the body came from this earth, whereas the human soul was implanted by God. At the beginning of the modern era, the philosopher Rene Descartes finalized this dualistic ontology and made it the basis of modern philosophical and scientific thinking. Descartes's philosophical model became known

---

35 David Skrbina, *Panpsychism in the West* (Cambridge, MA: MIT Press, 2005).

as Cartesian dualism. It postulated that the mind was a thinking thing (*res cogitans*) and, as such, the product of a separate realm of pure consciousness, which was assumed to be divine. The body was an extended thing (*res extensa*). Closely watched by a still powerful church, Descartes was happy to gain complete control over the extended material world to be used by the fledgling young scientists. The church, on the other hand, remained in firm control of the soul and everything mental. Cartesian dualism became the norm in the increasingly scientific world of the West. The result was what became known as the machine universe—a totally predictable world in which everything ticked with the precision of clockwork. Supremely sanctioned by Isaac Newton's famous laws of motion, the clockwork universe was the seemingly unshakable explanation of everything in the universe. The uncertainties of panpsychism did not fit in with the precision of this machine.

During the course of the twentieth century, a number of scientific discoveries were made that put in question the veracity of a machine universe and of a strict determinism in nature. But as we all know scientific paradigms are slow to change.

The philosophy of panpsychism touches the center of who we are. We are creatures who have, or at least believe to have, an intuitive experience of our own self. As Descartes pointed out, this experience of one's own thinking self is the only complete certainty we have. It is a truth we can be sure of, not by deduction, but by intuitive awareness. "I think, therefore I am." While this is true for each one of us, when it comes to defining what this thinking self is, we encounter deep problems. And while I am sure that I have a mind, this does not prove a priori that everyone else also has a mind, and much less that inanimate objects have a mind. As Skrbina correctly points out, this intuitive awareness of our own self led the unprejudiced primal mind to assume that all things have an inner mental agent. Panpsychism after all is in many ways similar to the ancient belief that all things are alive and have spirit, which is called "animism". While mainstream scientists and Western philosophy followed the analytical model that placed the origin of mind and matter into two separate realms, modern panpsychists hold on to the intuitive belief of ontological unity of mind and matter they had inherited from ancient animism.

Can panpsychism ever be tested? This is a difficult question. None of us has a crystal ball to know what might be possible in the future. But one thing is clear: should panpsychism, the view that all things have some kind of mentality, ever be proven, this would most likely look different from other empirical proofs that dominate science today. Current scientific theories have, as of yet, no testable explanation for subjectivity, or subjective experience.

Perhaps aware of the possibility that a traditional analytical proof for the validity of panpsychism may never be forthcoming, Skrbina calls panpsychism not a theory, but a metatheory. What he means with this is that panpsychism "does not in general give an account of the precise nature of mind, nor how it is related to material things."[36] Panpsychism, Skrbina contends, is "a theory about theories, a framework which effectively says, 'however mind is to be conceived, it applies (in some sense) to all things.'"[37] It is a theory in search of testable hypothesis, awaiting better tools of measurement. Perhaps a real proof will become available at some later time. Many scholars are currently reviewing first person phenomena for their suitability to scientific scrutiny. New methodologies and whole new sciences have already been invented and developed.

Accepting panpsychism could offer a new solution to old problems and facilitate a more acceptable solution to others. Carl Sagan once said the science that can do with the least number of miracles is the best. While for some, panpsychism may appear to pose a miracle itself, the reader will find that it is not as miraculous as a mechanical universe full of inanimate objects giving rise to life with all its self-determination.

Current theories in physics suggest that a large part of the universe, well over 99 percent, consists of insentient particles. These particles tumble around according to the second law of thermodynamics, guided only by the laws of motion, physical constants, and the second law, which dictates the slow decay of any order or diversity of energy

36 Ibid., 6.
37 Ibidem.

distribution established by the force of the initial big-bang event. Nevertheless, from these tumbling particles, somewhat miraculously a certain order develops, which we call life. Many cosmologists realize how improbable such a proposition is. The physicist Brandon Carter first called it the anthropic principle. Physicist Roger Penrose summarized it as a relationship between physical constants like the gravitational constant the mass of the proton, the age of the universe, and the speed of light. All these values must be "just right" for the existence of life on earth at this moment. This enormous improbability lead some to speculate upon the necessity of divine creation. Others saw it necessary to invent multiple universes to account for such improbability. According to these multiverse theories, our universe is only the last in a long line of universes—each one lasting for several billion years. All this is conjectured to allow enough time for probability to achieve its tremendous and unlikely task to create a universe with the perfect physical characteristics that could allow you and me to exist. The sheer complexity of nature, physical constants, and the life it brought forth is so unbelievable that some mathematicians have to assume billions and billions of years through multiple universes to even consider a statistical probability that could achieve this task through basic physical interactions. With the acceptance of panpsychism, self-organization and collective orchestration can offer a new understanding of nature. It makes the acceptance of evolutionary principles, which so far are reserved for living things alone, plausible for all of nature. Perhaps through the force of collective orchestration, matter "evolved" to fit the physical constants of this universe?

Although I find panpsychism extremely appealing, it may not entirely be necessary for the acceptance of collective orchestration. Collective orchestration is a viable theory even if panpsychism is not part of the explanation. By applying the second law of thermodynamics and evolutionary principles, we can show the practice and necessity for cooperation in the universe.

Furthermore, panpsychism allows for the whole universe to be conceptualized as one gigantic organism similar to the way planet earth has been conceived as one organic system. This became known as the so-called *gaia* principle. The mathematics of dimensions would suggest that our universe exists as an interactive organism, residing in an altogether different dimension. Because, as human beings we only grasp three dimensions, we have an incomplete picture of the universe and would be unable to experience these other dimensions, much like Edwin Abbott Abbott's flatlanders. We know our universe only where it intersects with the three dimensions.

CHAPTER 12

Superorganism

• • •

ACCORDING TO THE EIGHTEENTH-CENTURY PRE-DARWIN-IAN world, animals were brutish, primitive, and incapable of selfless acts. Real cooperation was possible only among civilized human beings. The idea of altruism, the compassionate caring for others, especially of those who were not one's relative, family, or tribe was believed to be the beneficial result of religion or culture. Without religion or culture, human beings would be just like animals—selfish and brutish.

At the middle of the nineteenth century, when Darwin wrote his famous books, the scientific view of the world was pessimistic. A clockwork universe followed eternal laws without mercy. Descartes had declared animals to be nothing but lifeless machines. Their suffering was of little significance. As religions and cultures declined, and revolutions and wars devastated Europe, people lost sight of the more noble aspirations spread by the Enlightenment. In the pitiless backdrop of the Industrial Revolution, the simple message of a crude materialist philosophy took hold. Human beings are not much more than animals, one pitted against the other, each ready to kill in order to survive. Altruism, selflessness, love, and cooperation were anomalies, the exception to the rule. The norm was individuals fighting for survival. Only a strong leader and tough laws could save human beings from themselves.

In this atmosphere of cultural despair, Darwin's message resonated well. Darwin's emphasis on random chance, competition, and fight for survival replaced Aristotle's teleological philosophy. With his belief in a grand design, Aristotle seemed unjustifiably optimistic. No longer could war-weary people accept it as true that the gods had endowed everything in the world with an ultimate purpose, and the best anybody can do is to live out that purpose. Now Darwin seemed to say that the only thing needed for creation to unfold was random chance and a crude instinct to survive. Social Darwinism turned an extreme version of Darwin's message into social dogma.

On closer reading, however, it becomes evident that Darwin saw nature more differentiated. He realized that groups of individuals cooperated not only among human beings but also in the animal world. Human beings and sometimes even animals seemed to be willing to sacrifice their life for the common good. Darwin realized that this kind of behavior was quite contrary to his theory and supposedly called such altruism "a stubborn anomaly of nature." He even believed this to be "a fatal challenge to his theory of natural selection."[38] To this day, many scientists consider cooperation as irrational or an invasion of something strange.[39]

According to Darwin's theory of evolution, things in nature change not to fulfill a predetermined purpose, but because the sense of survival of each individual dictates adaptation to new situations in the environment or risk perishing. Challenges in the environment may be the result of random mutation or of other individuals' adaptive actions. The survival instinct in each individual is stronger than anything else. Today, it is evident that traditional interpretations of Darwin's theory do not adequately address the obvious occurrence of cooperation and of individual sacrifice for the sake of the group.

Darwin, of course, realized the existence of selflessness, sacrifice, and cooperation in the human and animal world. It was most evident among insects. He did not know what to make of it. For over a century, the scientific community didn't recognize this as anything more than a fluke. For a century, this problem remained one of the hottest debated questions in evolutionary biology. In the middle of the twentieth century, William Hamilton developed what became known as "inclusive fitness theory." This quickly became the standard for evolutionary science. It

---

38 *The New Yorker*, March 5, 2012.
39 Michael Argyle, *Cooperation: The Basis of Sociability* (London: Routledge, 1991); Feng Zhang and Cang Hui, *Eco-Evolutionary Feedback and the Invasion of Cooperation in Prisoner's Dilemma* Games.

goes something like this: genes for altruism could evolve if the benefit of an action exceeded the cost to the individual once relatedness was taken into account. E. O. Wilson, a famous evolutionary biologist, first embraced this explanation and only recently changed his mind.

In a joint publication Wilson and his colleague Hölldobler published a much-discussed book,[40] *The Super-Organism, The Beauty, Elegance, and Strangeness of Insect Societies.* The book deals with the strange occurrence of so-called eusocial insects. These insects live in large colonies and cooperate on many tasks. These eusocial insects are among the most successful creatures on earth, far more successful than even human beings. Contrary to Darwin's theory, the selfish characteristics of these insects are far from predominant. Rather, they live in large cooperating communities. These communities function so well together that it often appears that they operate as one single organism. Darwin named such a cooperative body a superorganism, but could ultimately not make much sense of it. Darwin did not recognize the recurring pattern of such hive-cooperation. Acting as one body is the goal of cooperation and the ultimate outcome of collective orchestration. Not only does the individual in a hive or swarm not know the whole picture, at the height of the newly established body, each individual becomes a constituent part of a larger organism,

---

40 Bert Hölldobler and E. O. Wilson, *The Super-Organism, The Beauty, Elegance, and Strangeness of Insect Societies* (New York: Norton, 2009).

a superorganism. Individually each ant is "dumb." But as a swarm, they are rather smart. Recall what Jeremy England learned about the intelligence of individual ants.

"I used to think ants knew what they were doing. The ones marching across my kitchen counter looked so confident, I just figured they had a plan, knew where they were going and what needed to be done. How else could ants organize highways, build elaborate nests, stage epic raids, and do all the other things ants do?

Turns out I was wrong. Ants aren't clever little engineers, architects, or warriors after all—at least not as individuals. When it comes to deciding what to do next, most ants don't have a clue."[41]

Deborah M. Gordon, a biologist at Stanford University agrees: "If you watch an ant try to accomplish something, you'll be impressed by how inept it is," she says.

Each individual ant is part of a whole and receives instruction from the whole, effectively becoming part of a superorganism. Creating a superorganism is the essence of collective orchestration. According to Hölldobler, a superorganism is "a group of individuals in which the members are so tightly organized and tightly united that they develop traits characteristic of an ordinary organism."[42]

---

41 Jeremy England, "Swarm Intelligence," National Geographic Journal, November 2015.

42 Bert Hölldobler lecture presented as part of the *Darwin Distinguished Lecture Series* at the Phoenix Botanical Gardens. These events are sponsored by Arizona State University.

What are the traits that mark a superorganism? How are they evident in a superorganism? We said earlier that an organism is a system with many subsystems. System-wide communication would without doubt be an important characteristic of a system. This is the case with the cells in my body. Each cell receives a message when sickness befalls some part of my body. This knowledge may be disseminated through chemicals and through nerve connections. Another characteristic would be a sense of shared history. Each part of my body "remembers" some sort of collective past of the rest of my body. This explains how in a beehive an individual bee can have the knowledge about a feeding ground that extends beyond the life span of an individual bee. These are two important characteristics that superorganism have in common with ordinary systems.

For an individual, bee or ant, or whatever, becoming part of a larger whole means giving up a good part of independence. The boundaries of the self become blurred. To say it more philosophically: the individual must self-transcend.

In his TED talk, "Religion, evolution, and the power of self-transcendence," social psychologist Jonathan Haidt[43] asks whether this capacity for self-transcendence has an evolutionary advantage or whether it is an evolutionary mistake. He asks "how could it ever be good to lose ourselves? How could it ever be adaptive for an

---

43 Haidt, "Religion, Evolution and the Power of Self–Transcendence."

organism to overcome self-interest?" This line of thinking leads to the important and much-discussed question of group selection. According to Haidt:

> "The main argument against group selection has always been that, well sure, it would be nice to have a group of cooperators, but as soon as you have a group of cooperators, they are just going to be taken over by free-riders, individuals that are going to exploit the hard work of the others."

Haidt uses a simple computer simulation to illustrate this point. Let us assume, a group has evolved into cooperators. "They graze, they defend each other, they generate wealth…they grow, and when they have doubled in size, you see them split, and that's how they reproduce and the population grows." All would be good, but this peace is destroyed by a new group, whose members have a mutated gene, that directs them to be selfish. The simulation shows that "in short order, the cooperators are done for. The free-riders have taken over." Haidt concludes that "if a group cannot solve the free-rider problem then it cannot reap the benefits of cooperation and group selection cannot get started."

Nature's solution to this problem is, according to Haidt, the creation of a superorganism. He uses the cooperation between mitochondria and the cell nucleus as an example: "they used to be free-living bacteria and they came together and became a super-organism. Somehow or other—maybe one swallowed another; we'll never know exactly why—but once they got a membrane around them, they were all in the same membrane, now all the wealth--created by the division of labor, all the greatness created by cooperation, stays locked inside the membrane and we've got a super-organism."

Haidt ends the simulation game by pointing out that "a superorganism can basically take what it wants. It's so big and powerful and efficient that it can take resources from...the defectors, the cheaters."

But what exactly makes an organism "so big and powerful and efficient"? What can insulate an organism in such a way that others can no longer influence it in a negative way? The answer is complexity and dimension. A superorganism has reached a level of complexity that makes it inaccessible, perhaps even invisible, to the lower-grade organism. The secret lies in the dimension, which is the degree of freedom the new superorganism has reached when compared with the other organisms.

In Edwin Abbott Abbott's novel *Flatlanders* (1884), the flatlanders have two-dimensional awareness. They live in only two dimensions. We can use their perspective to visualize the power of dimensions. These flatlanders would find it impossible to even imagine a creature of the third dimension. An organism living in the third dimension would be truly inaccessible to them. From the perspective of the flatlander, a three-dimensional being would remain mysterious. When European painters in the late Middle Ages discovered the third dimension, it added a whole new level of complexity to their art. Artists had found the third dimension. The creation of a superorganism involves gaining a new degree of freedom, a new dimension. With the escape to a higher dimension, nature has found the most effective way to insulate an organism from free riders and in the process, create complexity.

Where once Darwin found himself utterly perplexed by what appeared to be the unexplained emergence of superorganism, we now find ourselves with the theories and tools necessary to understand them. Jeremy England's Dissipation-Driven Adaptation theory demonstrates how energy use within a system can lead to increased complexity, despite the pull of entropy. The equations he developed to support this theory not only demonstrate how the sudden emergence of complex systems could arise, but suggests this move towards complexity is exactly what we should expect from systems where energy acquisition is a survival advantage. When we further apply Collective Orchestration to the investigation of complexity's role in the energy efficiency of systems we begin to see how it could also explain Wilson and Hölldobler's eusocial insects or biologist Deborah M. Gordon ant swarm intelligence. These systems are achieving increased complexity because their constituent parts are cooperating agents and not independent actors driven by individual survival. An individual ant could track off on its own and not share any of its food collection efforts with other ants, but this ant's likelihood of survival would decrease drastically. However, as part of the group, the shared system becomes the main mechanism of survival for which individuals are willing to sacrifice themselves to maintain. From here, the transition to superorganism is yet again a leap in survivability. As the superorganism imparts the group with a new degree of freedom, they leave behind freerides and other competition for energy recourses.

CHAPTER 13

From Symmetry to Complexity

• • •

As a child, I admired flocks of geese flying in perfect V-formation. I often wondered why geese would do this. What was the deeper reason behind it? Later I learned that they would fly like this for practical reasons. They had learned that flying in this formation saved energy for the whole group and that only the goose flying in front had to struggle harder. The geese rotated leadership so that all the members of the group could participate in the saving of energy. How had these animals learned that energy conservation was more efficiently done by a cooperating group? It became clear to me that energy conservation was part of the ongoing struggle to survive. Any system that is not concerned with the preservation of energy will soon succumb to natural selection and ultimately to the throngs of the second law of thermodynamics. Everything is in this continuous struggle to preserve order and avoid death. In the ongoing struggle to survive, those geese that have learned to cooperate had a better chance of survival than those that tried to migrate on their own.

Could the same be true for systems on a smaller scale? Going into relationships with each other and lining up as a group in a geometrical formation seems to be a common occurrence among small particles such as electrons and atoms. The Bose-Einstein condensate provides a good example. When plenty of energy is available, atoms tend to form gases. They tumble around freely as if they were oblivious of each other. If energy is withdrawn, that means when atoms are cooled down, they act as a group and organize in a more uniform fashion. Withdraw even more energy, and they become a solid matter. The scientist Bose predicted that if matter were cooled down to almost absolute zero, something very strange would happen. The formerly free-roaming atoms would all draw together and occupy the same space, as if they were one single atom. Bose proposed this in a paper to Einstein. This material state, which was later verified, is named the Bose-Einstein condensate. The phenomenon has also been called a super-atom. Though some might find it strange to call this activity cooperation, in a sense that's exactly what it is. On lacking sufficient energy to survive by themselves, many atoms come together and survive as one giant super-atom. When faced with a lack of energy, individual atoms cooperate in order to survive.

Synchronization occurs in the form of symmetry. Just as it was in the case of the geese, it is the symmetrical arrangement of individual members that preserves energy. Deacon speaks about "this curious geometry of chance," which he describes as "a relational property." It is the formal constellation of individual systems in space that causes spontaneous change and leads to energy conservation. Not unlike a flock of geese, small particles move into symmetrical formation to save energy. These changes in contrast are spontaneous, coming from the self-organizing power within each "living" system.

The question remains how does collective orchestration create complexity? Simply said, complexity is a way for a cooperating group to avoid being swallowed up by free riders. Jonathan Haidt pointed out that in any cooperating group, free riders may develop and threaten or even destroy the group. A cooperating group can avoid this destruction by creating a more complex superorganism. In the first part of collective orchestration, individuals come together and synchronize their actions. The individuals begin to act as one. On rare occasions, the newly formed collective does not spread apart but remains together. When this happens, a so-called threshold situation occurs. Now the cycle is complete. This completion of the cycle can lead to the creation of a whole new species at a completely new level, with increased degrees of freedom.

CHAPTER 14

The Evolutionary Use of Ecstasy

● ● ●

IN HIS TED TALK, "RELIGION, evolution, and the power of self-transcendence," social psychologist Jonathan Haidt proposed to imagine the mind as a house with many rooms. Under certain conditions, he says, the familiar self can just melt away. "...it's as though a doorway appears from out of nowhere and it opens onto a staircase. We climb the staircase and experience a state of altered consciousness."

Haidt proceeds to describe a number of conditions in which this altered state of consciousness may occur. He cites William James who reports the incident of a young man who had the experience of meeting Jesus. In the process of climbing the staircase, he leaves all selfishness behind and becomes loving and forgiving. Haidt recounts a variety of other situations in which the familiar self is transcended through the cooperation of a group. For warriors, the experience of the communal effort in battle produces this kind of ecstasy. The familiar self disappears, "the individual fate loses its self-importance. I believe that it is nothing else but the assurance of immortality that makes self-sacrifice at these moments so relatively easy."[44]

What Haidt describes here is at the center of collective orchestration. A group of disconnected individuals, who are driven by a common goal, band together. They eventually achieve a high level of cooperation and eventually begin to act as if they were only one body. They have effectively turned into a superorganism. The group has transformed from a disorganized crowd to a highly organized one. This process of many becoming one usually produces a heightened state of awareness among the participating individuals, an ecstasy that melts the individual borders of self away and makes self-sacrifice possible or even attractive.

This may address the question on why there is self-sacrifice in nature from a new angle. The loss of self is a central theme in Buddhist teaching. Practitioners learn various techniques designed to silence the self. The human self is seen as the cause for feelings of division and separation from the universe. Once the divisive character of the self is overcome, the mind is free to experience total connectedness. Individuality is not only perceived as an illusion, it is the source of suffering. What from the Buddhist perspective appears as a defect, an undesirable condition to be overcome, the Western mind celebrates as

---

44 Glenn Gray as quoted by Haidt in "Religion, Evolution and the Power of Self-Transcendence."

one of its greatest achievements. Loss of self is often seen as an unwelcome pathology.

The evolutionary purpose of ecstasy has long been in question, as Haidt pointed out. It is rather often seen as an evolutionary mistake especially when we consider the hysteria of large crowds or the self-destructive activities of battling armies in war. On the other hand, it would seem that evolution does not allow for such a ubiquitous occurrence without having a constructive purpose. We can certainly find plenty of cases in which ecstasy is used for a positive reason. Haidt points at self-transcendence in nature and the loss of self in techno raves as well as dancing, spinning, and circling.

Remarkably many religions were initially conceived by an ecstatic experience of its founder. Ecstatic mystical experiences have as a centerpiece the sensation of complete oneness. Could it be that our journey through this world is completed or perhaps continued with the submersion of our individual selves into yet a higher-level unity?

Let us take a closer look. In the process of many organisms becoming one superorganism, there are a number of transitional phases. Initially, many individual organisms—each separately—assert their identity in the ongoing struggle to survive. As individual systems begin to synchronize their actions, and as many become one, the survival needs of the whole takes precedent over each individual's survival. Just as in a human body where thousands of cells die off and are replenished by the system as needed, so are individual members of a newly created superorganism subject to replacement if the new system requires this. For example, in the case of the dictyostelium, about 30 percent of individual amoeba die off and build the outer skin of the new organism. Individuals are "sacrificed" to avoid death of the whole. What appears as self-sacrifice from the vantage point of the lower-level individual is part of the routine of maintaining life from the higher perspective of the whole.

This sacrifice appears to be connected to an ecstatic frenzy or an altered state of mind. In this frenzy, the individual is willing to accept the sacrifice. This makes ecstasy a necessary occurrence in nature.

The process of collective orchestration begins with a period of heightened sensual awareness. As Haidt puts it, "I" passes insensibly into a "we," "my" becomes "our." In the metaphor Haidt uses, ecstatic selves are yanked up a staircase. As all individuals act as one, they can collectively escape to a higher dimension. In Haidt's words:

> "Bradley's petty, moralistic self just dies on the way up the staircase. And on this higher level he becomes loving and forgiving. The world's many religions have found so many ways to help people climb the staircase. Some shut down the self by using meditation. Others use psychedelic drugs."

Haidt cites a sixteenth-century Aztec scroll showing a man eating a psilocybin mushroom. At the same moment, he gets "yanked up the staircase by a god." Haidt continues: "Others use dancing, spinning and circling to promote self-transcendence. But you don't need a religion to get you through the staircase. Lots of people find self-transcendence in nature. Others overcome their self at raves."

Ordinarily every individual obeys the law of self-preservation. In the course of collective orchestration, however, individuals give up their natural boundaries so that many can become one, but on a higher and more complex level. The willingness of each individual to give up its boundaries is facilitated through this collective frenzy, whether this is in the form of vibration, whirl, or right-out ecstasy. This delving into an altered state of mind enables individuals to sacrifice their selves and, in extreme cases, to give up their lives as the amoeba do in dictyostelium or an individual soldier in war.

Each superorganism is the result of the sacrifice of many less complex systems. Each cell gives up the boundaries of its self in order to become part of a larger self. The accompanying ecstasy makes this sacrifice acceptable. Self-organization assures the voluntary character of the process.

CHAPTER 15

Absolute Nothingness—An Occasion of Experience

• • •

I GOT MY FIRST INSIGHT into absolute nothingness when I was barely a teenager. From physics we know that nothingness, or empty space, cannot be achieved. It appears to be impossible for a physicist to create a perfect vacuum. Out of seemingly nowhere, strange particles spring into being and populate what we believed is empty space. Thus, from a physicist's point of view, the concept of absolute nothingness seems impossible, even absurd. And yet, the idea has occupied the attention of many great minds, from philosophers, to mystiques, to poets. One thing is clear, absolute nothing cannot be something physical. In fact, it must be the negation of everything physical.

As a young man growing up in Germany, I attended a gymnasium, which is the equivalent of an American high school. My instructor in German literature, Dr. Walter Gerstberger, was a rather peculiar man. Small in stature, always wearing a wrinkled suit, this soft-spoken man was supposed to teach us German literature. Instead, he lectured us on the intricacies of the universe. Gerstberger often talked to us youngsters about his personal experience as a mystic. Most of the time, we understood little or nothing of what he was saying. Yet, his lectures left a deep impression on me—something that would stay with me for the rest of my life. Frequently, he quoted Goethe's *Faust* (the only piece of literature he seemed to value). With his hero, the always-inquisitive Faust, Gerstberger took us down into the depth of all being, a place so unusual, so inhospitable that words were not able to express it. Goethe called this secretive location "the place of the mothers", the place of original creation that was located somewhere before space and time. It seemed like a place so terrifying, not even the devil (called Mephistopheles) was able to approach it. Faust had to venture there alone. The place of the mothers is beyond reflection. It must be experienced as the eternal now. Even expressed as a thought, it loses its originality.

Gerstberger claimed to have been there. The mind, he said, stripped of everything that is real, can experience this ultimate reality—absolute nothingness—which at the same time is absolute unity and absolute identity. All this, Gerstberger said, was mathematically expressed by the symbol zero. But if this wasn't enough, zero—this absolute nothing—was able to identify its own identity. What in the world was that supposed to mean? It was a thought, but it seemed impossible to grasp.

Gerstberger frequently used the image of a mirror ball, a hollow sphere. "Put your mind into the center, what would you see?" he asked; his small eyes sparkling. "Nothing, of course, nothing." You would find yourself in absolute darkness. Now if your mind could light a match, what would you see then? Infinite light—an explosion of light all around you. The feeble light of the match would be mirrored into the infinite depth of the mirror mirroring itself. But how does the mind get inside the mirror ball? Where does the first spark of light come from? These were questions I could not help asking throughout my life.

While Western philosophy is largely concerned with conceptualizing and making the world understandable, the goal of Eastern wisdom is to create experiences. Nirvana, the experience of absolute emptiness or nothingness, is rarely achieved. Nothingness is the eternal now.

Living in the present has become somewhat of a fashion. It is the experience of now-time that has created a bridge between the East and the West. In Eastern thinking, being in the now was the goal. The Western mind, overloaded with concepts, had to come full circle to embrace the experience, specifically the experience of now. The Hegelian model of a living dialectic is perhaps the most adequate bridge, a bridge that extends across the abyss to the other side of which we can only hope it exists, as Merleau-Ponty once put it.

To be sure, this is a very refined understanding of the Hegelian triad. It is expressed in Hegel's aesthetics and nowhere else. Goethe again expressed it best when he had his Faust respond to Mephistopheles's skepticism. When Mephistopheles cautions: "You will see nothing in all eternity." Faust responds: "In your Nothing, I hope to find my All." It is a matter of perspective—and faith. Only the left-Hegelians understood Hegel's dialectic in quite that way. Politically, it became the cornerstone of permanent revolution; aesthetically, it found its way into the Frankfurt School philosophy, more specifically in Adorno's negative dialectic.

The Spirit, the Absolute, God, according to Adorno, had withdrawn "into the micrology of things." The Absolute could not be found by reflective abstraction. Kant had proven this impossible. Adorno believed that modern man can only meet the Absolute, or fragments of it, in the depth of things and after clearing the mind of all concepts. This is obviously also the goal of nirvana, the Buddhist's experience of the Absolute. This experience is only possible in the now. Any formulation of it in words, even thoughts, ends in loss. Goethe, who realized this, used an appropriate image. The mind is like the head of Medusa. Everything she puts her eye on turns into stone. Everything the mind touches, thinks, every thought is the opposite of the Absolute. Adorno's negative dialectic formulates nothingness by carving out an experiential place for it. The Absolute is, in the words of Alfred North Whitehead, nothing but "an occasion of experience." As that it is all.

CHAPTER 16

Complexity and the Arrow of Evolution

● ● ●

Something is complex when it has many parts. In this sense, collective orchestration creates complexity. On first sight, this seems counterintuitive. Is not the opposite the case? The process of collective orchestration is summed up by the simple phrase, *e pluribus unum*, which means "from many comes one." This is indeed the case when a group of individuals come together and make one individual, as in the example of the dictyostelium. The stalk created by this collection of one-cell organisms now possesses an intricate structure. The new body reflects the lower-level amoeba. Each of them now assumes a specific function. Recall that the stalk has a simple digestive system and a rudimentary brain, not to mention the thousands of amoebas that died in the process of forming the outer skin.

It is true that many have become one, but this new entity, if completely successful, has the ability to replicate itself and set the basis for a new population. In effect, the new organism may well become the birth mother of an entirely new species.

This adds an interesting level of development to evolution. This is a temporal advancement in the level of complexity. The second law of thermodynamics defined the arrow of time. Order, left to its own device, inevitably turns into disorder. By all practical means, this process is irreversible. An egg, once it is broken into the frying pan, looks unrealistic and weird when in a backward-running film the broken egg jumps out of the frying pan and the pieces come together reconstituting the unbroken egg. In classical physics, in contrast, all processes can theoretically be played backward without appearing strange. The fact that in classical physics all processes can be reversed resulted in the universe representing a machine. But this mechanical universe coming from the rigorous application of Newtonian mechanics proved deeply flawed. The second law made it clear that everything in this world is in a slow process of decay.

The evolution of life tells a different story. Living systems seem to steadily increase their complexity. This happens through the process of collective orchestration. Evolution, according to Darwin, clearly did away with Aristotle's notion that everything strives toward a defined end. Darwin concluded that there was only blind chance and competition, in which the better organism survived. While there was development, this advance did not follow any grant plan. According to Darwin, evolution is blind of purpose.

What happens if we add cooperation? Does this add an arrow to evolution? On the surface, there seems to be little difference. It is still true that the fittest system will survive. Only now we can say that any system that cooperates with others more effectively has a better chance at it. Nothing has changed here. But on closer analysis, collective orchestration adds a significant direction to the evolutionary process. As individual systems cooperate to survive, they collectively advance in the level of complexity; thus, while there is no horizontal goal for the arrow of evolution, there is vertical advance. Nature progressively advances to higher dimensions. To escape the destructive influence of free riders, nature creates organisms with increasingly higher degrees of complexity and freedom.

CHAPTER 17

Implications for Morality

• • •

WESTERN CIVILIZATION RESTS ON A surplus of guilt. It begins with the expulsion of the first human beings from paradise. As a result of eating fruit from the tree of knowledge, God punishes the first man and woman by closing the gates. It ends with the martyred Christ on the cross. Left to their own devices, human beings would have been helpless and lost.

A divine redeemer was needed to rescue humanity. So, the story goes, and it left human beings to the mercy of natural forces. Without direct intervention, there was no hope for humanity. Morality was part of the divine package that was handed down from heaven.

The time has come to change this narrative. The discovery of chaos theory marks the beginning of a new era. Out of chaos comes order. It appears that nature is not all hostile and alien to the human plight. We now understand that the conception of a mechanical universe was a grave mistake, born on the ruins of abhorrent injustices, great wars, and misguided philosophies. A hostile world was the price science paid for building its marvelous edifice not on the faint flames of the Enlightenment but on the back of ancient mythologies and stories.

But the direction was decided even long before the first human beings walked the earth. We will have to go back to the origin, not the scientific one, but the origin of everything, how ancient people conceived it. This story is called the "Myth of Origin," and a myth it is. There are many versions told in different cultures to explain the birth of humanity. All of them can be considered myths of origin. How these different cultures and civilizations expressed the coming to be of the universe and the birth of the first human beings presents a kaleidoscope of diversity.

When those ancient storytellers observed the world, they found two fundamental states seemingly opposing each other: chaos and order. Where they placed chaos within their origin myth played a decisive role. Was chaos a destructive force or was it benign and even creative? According to the Egyptian *Book of the Dead*, chaos created itself like a snake body out of nothingness. In Western mythologies, chaos is more often depicted as an opposing force, destructive and dangerous. The whole universe was believed to be chaotic, and an ordering force was placed outside of the universe. In Christian belief chaos was seen as an evil force, often identified with Satan. Nothing good could come from it. Once the whole universe was considered chaotic, that universe had to be dark and evil. It was no place to search for the roots of morality. A narrowly defined moral advice put great emphases on the negative aspects of human beings.

This focus on negativity carried over into the world of science. Even psychology was mostly used to uncover human shortcomings and defects. Today we know that this is not all psychology can do. Clearly, it is of value to make sick people healthy. But since the doctor's job was to cure sick people, often pathology was projected just to keep the good doctor busy. How much more satisfying would it be assisting healthy people to develop into the best they could be and to use scientific power to spread happiness among people?

Just such an endeavor is undertaken by a still small but growing number of forward-looking psychologists. Their work is known as positive psychology. Positive psychology is defined as "the scientific study of the strengths that enable individuals and communities to thrive. The field is founded on the belief that people want to lead meaningful and fulfilling lives, to cultivate what is best within themselves, and to enhance their experiences of love, work, and play."[45]

As Aristotle already stressed, morality is community based. Through the many laws and commandments, our offspring are guided to live in greater harmony with the community they grow up in. The essence of community is cooperation. This was still clear to Aristotle, but over time, community lost its appeal. As laws became

---

45 www.positivepsychology.org/Mission statement

increasingly more universal, philosophers put more emphasis on the individual than on community. A universal set of laws slowly replaced community law. Where in earlier times the worth of a person was defined by the community he or she belonged to, the emancipated individual of the Renaissance found value in the self. By the nineteenth century, the destruction of community was in full swing. Competition was the new name of the game. In such an atmosphere, cooperation seemed like a misplaced idea. To survive, you had to compete. This was understood as the new law of nature.

But as long as survival depended solely on competition, we find community and "natural" law squarely pitted against each other. This was the case under the Social Darwinian paradigm. Meaningful cooperation was seen as a fluke, a mere by-product of each individual's natural drive to secure offspring. Community was perceived as an almost useless endeavor we can easily do without. Nature was seen as selfish and violent.

Collective orchestration has shown that building and nourishing communities is an essential ingredient of the universe and everything it contains. Cooperation is necessary for systems to survive and flourish and eventually reaching levels of higher complexity. Among human beings, this attempt to preserve and secure community has been the seed for laws, commandments, and rules that eventually turned into the complex body codified in morality and ethics. None of this could be derived or learned from animals, even though they are our closest allies sharing the planet with us. The connection was broken.

Today, it is quite evident that animals show affection, caring, and even love in their daily life and often demonstrate quite a refined sense of justice. Especially animals living in large groups, which seem to follow distinct rules.

Why were these obvious traits among animals ignored for thousands of years? Serious research on animals in their natural habitat has only been conducted for the last century. For much of recorded history, animals were not considered suitable to support a foundation of morality. The Christian West took its directive from the biblical mandate to subdue the earth and all its creatures. Only those animals that were of use to a farmer for his exploits were deemed worthy. All others were simply called "wild". What a far cry from the relationship Native Americans shared with animals, which they considered to be their brothers and sisters. As far as the learned man in the West could see it, nothing good could come from animals. The human being ruled supreme.

At the center of morality is the question of how to establish ethical norms. What is considered ethically good? On the surface, the drive to cooperate and to build communities does not yield an ethical norm. Collective orchestration can create many different types of communities. Which ones deserve to be fostered and which ones should rather be opposed or should perish? Since cooperating communities must follow the law of natural selection, one could simply say that the better ones will survive and the bad ones will go under. But how long must we wait? Obviously, we can learn from history. Bad communities seem to disappear faster than good ones. The Thousand-Year Reich of the German fascists lasted a little over a decade. The mighty Soviet Union broke apart in less than a century. The democratic United States in contrast, though contentious at times, has lasted several centuries and is still going strong. For those who listen and learn, history can be a powerful teacher.

In the face of such American exceptionalism, is it time for our American Faust to deliver his soul to the devil? A country where everyone is permitted to live in freedom and pursue happiness—wouldn't this be the fulfillment of Faust's dream? Add to it the guarantee for each able-bodied person to find adequate employment—the utopian dream appears to be complete.

Not quite so fast. In Faust's vision, a strong communitarian spirit drives the utopian community. This is what American society is sorely lacking. We experience communitarian spirit often only in the wake of great disasters. Human beings band together when suffering, but they show their selfish side in times of peace and prosperity.

Cooperation is a basic principle in the universe. As we have seen, there is cooperation at every level. No longer must the universe be seen as entirely strange and hostile. Order can grow out of chaos. In the center of this organic

process are self-organizing individuals. This cooperation saves and preserves energy. Those who cooperate
fare better in the ongoing struggle to survive. Collective
orchestration is the process in which it unfolds.

A final note of importance worth stating here. The
evidence for collective orchestration is not an argument
for collectivist cultures over individualistic ones. It is a
mistake to assume the celebration of the individual is itself
a removal of the individual from group level systems. It is
not. Cultures that celebrate either the individual or the
collective only differ in their group behaviors. There are
profound group benefits to the technological innovation
and entrepreneurial spirit currently associated with cultures identified as individualistic. But that isn't to say that
cultures identified as collectivist don't also possess these
traits. As stated before, the benefits of one value today may
not be equally as beneficial in tomorrow's environment.
A diversity of options within a larger group to respond to
future challenges should be considered a benefit, even if
managing that diversity today is a challenge.

Additionally, the insights of collective orchestration
are not an argument against competition. Competition
can be a cooperative process. Professional sports for
instance are not simply a competition of who can win.
Before the game even starts, both teams agree on a set
of values and rules which are designed to maintain the
very system of competition they are choosing to compete
within. When teams are found to be in violation of these

rules, the norm is to punish them. That's because the system they are competing within has a greater value than any individual win. The legal system is another rule based system where competition is built into the base structure of certain courts. Known as an "adversarial system", these types of courts operate best when the prosecution and defense work at odds with each other to discover the truth of a crime. The assumption being that if one side is being dishonest or concealing a key fact, the opposing side will work diligently to reveal the deceit. When the system is honored, the adversarial approach can be very productive. However, not every participant respects that integrity of the legal system and may place more value in some other goal; be it as malicious as racism or as mundane as job promotion. Convincing all participants to value the integrity of the system, even an adversarial system, should be an integral design specification for any successful human system seeking to learn from the insights of collective orchestration.

CHAPTER 18

Responding to Free Riders

● ● ●

WHILE CONSIDERING THE IMPLICATIONS OF collective orchestration on human society, we inevitably turn to the issue of free riders and how we should respond to them. We do not need to look far for examples of human free riders taking advantage of cooperating groups while not reciprocating anything of value back to the group. It would be easy to see how collective orchestration could be used as an argument to punish free riders into forced participation, or even banish them from the group to prevent their coercive effects on collective resources. But this view is short sighted, lacking a long-term understanding of system survival and growth. To be clear, we are talking about our fellow humans here. Collective Orchestration is not a moral system. It is a theory which describes the systems that make up our universe. Describing them the way current trends in science have found them. However, if we choose to look for moral guidance by comparing the systems of human society to other natural systems, we are not clearly lead to such emotionally cold conclusions. In fact, analyzing natural systems with collective orchestration suggests just the opposite.

Take the example of a forest replanting effort just north of Frankfurt, Germany. After a rather severe storm had leveled a large section of the local forest, it was decided the recovery effort should use trees with a much stronger trunk and root system. An entire hillside was replanted with this stronger tree. Unfortunately, planning for the previous disaster is not always the most logical path. It didn't take long for a local bark disease to come through and wipe out these new trees. The stronger trees were no longer susceptible to windy storms, but a species of the physically smaller trees had provided a patch work of protection against the spread of the bark disease. This changing challenge to the health of the forest in turn shifted the position of strong bark tree species from group benefactor to free rider. This story is purposefully left vague to emphasize the difficulty in always identifying which individuals are going to be free riders. It is equally as difficult to know beyond a benefit of doubt that a free rider could never serve the greater good of the group. Not because they were forcibly changed to become beneficial, but because the challenges to the system changed in such a way that weaknesses become strengths.

With that said, free riders are not to be ignored. They are after all a drain on the resources of the group. If groups that collect and use their energy more efficiently are the ones that ultimately out-survive other groups, then efforts should be made by the group to reduce a free rider's drain on recourses. The catch 22 here is that such efforts themselves often cost the group energy. Possibly even more than the energy consumed by the free rider. We are after all, talking about human free riders. As mentioned before, Collective Orchestration does not guide us to a brutal cost-benefit assessment of each person's energy value to the group. Instead, we see study after study where previously burdensome group members became vital to the group when situational challenges changed. During cooperative dispersal of bacteria, discussed earlier in this book, any bacteria could serve as the sacrificial food source for the group. Initial value states of bacteria that transition into this food role do not need to be considered when assessing its new value to the group in a food scarce environment. The lesson here is that any member of the group can transition into beneficial roles; and not that any human can become useful simply by turning them into soylent green[46].

In the first half of the twentieth century, the US military calculated that the growth of newly emerging officer positions like airplane pilots and tank commanders could not be filled by the previous method of selecting officers from among college educated, well-to-do families. The military began employing psychologist to devise methods

---

46 Soylent Green was the subject of a 1973 science fiction film where recycled humans were turned into food.

of filtering all new recruits into appropriate training for any given military position, regardless of family lineage or previous experience. Today, all levels of leadership in the US military are comprised of a diverse background of individuals. People who are both productive and unproductive can enter the same training program and learn to serve their military unit effectively. To emphasize, the US military training programs can transition previously unproductive members of society into a productive group member. All that's needed is the will to enlist and commitment to their program. Of course, the will to enlist is not for everyone, but it does demonstrate the difficulty in simply disregarding current free riders as burdens, lacking redemptive qualities. Nor does being a productive member of a group prevent one from becoming a free rider in future environments.

Thanks to advances in battlefield medicine, more soldiers are surviving the physical wounds of traumatic combat. These men and women felt the call to serve the collective will of their nation. Unfortunately, this advancement in battlefield medicine also means a growing number of veterans are suffering from the mental wounds of combat. Where once a soldier was called to serve their nation, these mental wounds are making some returned veterans feel unable to transition back into a productive civilian life. A naïve view of unproductive people as lazy and unwilling to participate cannot so easily cast these veterans into that caricature of free riders. After all, these men and women clearly answered the call to serve the group. The examples of collective orchestration we have from nature not only allow for, but demonstrate the success of both compassion and the collective sharing of burdens. A suffering veteran today could very likely become a productive computer engineer tomorrow, if the group shares the burden of rectifying the obstacles in her way.

Free ridership is a situation bound assessment. Group success is a balance between maximizing current energy resources and maintaining optimal future flexibility. Having both the energy and the freedom of choice to respond to the unknown challenges of tomorrow. Far from conflicting with modern human ethics, this outcomes-perspective mirrors the most optimistic of human morals. From public schools to international trade of goods and ideas, the current state of human ethical discourse is filled with calls for bringing more of us to a level of global participation and engagement. There are many situations where the state of nature is in direct conflict with modern human ethics. There is no absolute mandate in human ethics that demands our ethics be in harmony with all natural systems, like viral infections. Collective orchestration is not one of those conflicting natural theories.

Examples of collective orchestration demonstrate that groups can successfully overcome the burdens of free riders. Unfortunately, calls to deal with the burdens of human free riders comes in the form of protecting the group from "them". However, throughout nature, we see that productive inclusion of free riders back into the group helps improve future flexibly to respond to unknown challenges. This inclusion does not need to come in the form of forced participation, but in appreciation for an as-of-yet undetermined value the free rider may provide; to themselves and to the group. Countries where prisoners are the least likely to return to crime after prison, are countries like Norway where the community takes responsibility for rehabilitating prisoners. Countries that chose to spend the extra recourses necessary to rehabilitate prisoners often are small monocultures. Meaning prisoners tend to look like everyone else and come from similar backgrounds. This makes it easier for the group as whole to see prisoners as having been forced into crime by unlucky circumstances instead of being inherently evil and unredeemable.

Collective orchestrations may not be a moral guide, but it does offer us a plethora of examples in nature where groups working together can solve the problem of free riders without resorting to violence or expulsion. A key trait necessary for such a path within human groups would be to see free riders as inherently deserving of participation in your group's shared rewards. The benefits of such a perspective was demonstrated above with our examples of group complexity increasing to offer more diverse options for responding to threats. Additionally, energy not spent on resisting free riders is energy saved for long term group survival.

Research in psychology demonstrates that our base instinct is to divide the humans we meet into in-groups and out-groups. Those who get to participate in the benefits of being in our group, and those who do not. This research also demonstrates how arbitrary these distinctions are and how easily they can change if given the right cause. A classroom full of diverse students can be told that people who wear glasses are more intelligent and will get a higher grade in the class, so they will all get to sit up front. Almost instantly, those without glasses will likely form an in-group bond to challenge this dictate as a group. The eye glass wearers may also find group cooperation in their new found shared self-worth. Bonds like these are fun for class demonstrations, but aren't likely to last long, especially when it's reviled that there is no such research. However, more robust studies show that we divide ourselves mostly over these types of arbitrary differences.

Collective orchestration offers a clear value for learning how to increase the size of our personal in-group. By creating a more diverse wealth of relationships, we create a wider range of options for dealing with future challenges. In truth, this creates additional challenges, but also offers us access to resources we couldn't possibly maximize within our immediate social networks. Research into the wisdom of crowds shows that, on certain tasks, skillsets held by the masses can very often outperform the skill-sets of individual technicians. Companies like Google and Netflix have found time and again they can improve their company's performance simply by releasing their most valued computer code to anyone who wants to take a look. A famous Netflix challenge even found substantial improvement to the video recommendation algorithm within hours of releasing it to the public in what's called an open source challenge. To emphasize this power of the masses, Netflix had hired some of the best software engineers in the world to work on this algorithm for several years. And within hours of expanding the number of people working on the algorithm to the size of anyone willing to take a look, they increase efficiency beyond what the group of experts had accomplished.

We are born sorting each other into in-groups and out-groups. People perceived as a threat to our shared resources often get categorizes as out-groups, "others". But these categorizations are arbitrary. Collective orchestration demonstrates real value in expanding the size of our in-groups. Each new individual we learn to cooperate with, is an additional resource gained and future threat diminished. Additionally, using group resources to turn free riders into productive group members, will ultimately increase the complexity of our group and better prepare us for the challenges of tomorrow.

CHAPTER 19

The End to a Story's Introduction

● ● ●

AND SO, OUR STORY HAS been told. At some point, all systems require new energy to survive. Without new energy, entropy will ultimately deplete the system's usable energy. Systems that collect energy more productively and consume it more efficiently, out last comparable systems with inferior energy management. Complexity is the ever-improving tool that systems employ to maximize their energy reserves. With each increased degree of complexity comes the possibility of an increased degree of freedom for the system to respond to energy challenges. Ultimately leading to increased survival of the system. This ever-increasing complexity within a system of cooperating individuals has the potential to permanently change the collective into a higher order individual, a superorganism. From this new state of being, the system escapes previous challenges to its energy reserves, and the process continues. The higher order individual will begin to interact with other superorganisms of its kind, at which point they too will discover ways to cooperate for mutual benefit. This is the process of collective orchestration.

Our story begs the exciting question: what is the next level of human complexity? This is the type of question that fills the dreams of science fiction writers. Are we bound to become a superorganism? Science fiction writers have been exploring this theme since the early days of science fiction publishing. Some of the possibilities laid out in science fiction are fun to explore, while others are cautionary. The following examples were chosen for their popular culture appeal, and not due to origins of the concepts they explore.

In the Star Trek universe, we see a modestly integrated human collective within starships. Individuals dedicated to a shared cause and living in tandem onboard starships. They travel the universe with the unified mission of exploration. Individuals move about the ship like a mammalian circulatory system. The ship's computers operating like its nervous system. This may be the least integrated superorganism in science fiction. Some may even argue it's not a superorganism at all. However, from the outside perspective, the humans onboard those starships are bound to each other, as well as the ship they reside on. Their entire existence is intimately interwoven into a single mission. Not unlike the submarine crews of today's modern navies. Individuals may move in and out of the system, but the system is now in place and will live on. From here we can ask, is the system alone a superorganism? How we go about determining the existence of superorganisms is a challenge. There likely won't be a single point where integration shifts from individuals to a new collective superorganism. There may not even be just one shift or one type of superorganism. A starship may or may not continue to increase in complexity and interconnectedness. Similarly, we can ask, is the dictyostelium a group of individuals working together or a collective already in transition to superorganism?

The Star Trek universe also offers an incontrovertible human based superorganism in the Borg. A collective of individuals from different species who've given themselves over completely to the hive mind. Each individual is connected to the other through advanced cybernetics, all sharing one collective consciousness. The individual is lost to the whole and a single conscious experience emerges. The Borg of Star Trek is the ultimate science fiction cautionary tale of cybernetic integration. Currently, scientists have only begun to theorize about the nature of consciousness. So, while science fiction is filled with stories about transferring our consciousness into other objects, at present, we don't even know what would be the thing that gets transferred. None-the-less, the Star Trek Borg is a scary proposition. As our internet becomes wearable, and then permanently attached in the form of cybernetics, the individual might get lost to the collective. While such a radical loss of autonomy seems frightening, it would likely happen slowly over an extremely long span of time.

Even now, I am all too happy to lose certain elements of autonomy to the technologies I carry with me every day. The news I choose to read is selected by the up and down votes of the Reddit collective. I no longer remember phone numbers as I've ceded that boring task to my cell phone's computer. Nationwide surveys estimate that between 10% and 30% of new marriages in the US started as an arranged engagement…by an online algorithm. Not long ago the power of the soulmate myth was considered an obstacle to online dating. Now, Pew Research found that nearly 60% of US adults consider an algorithm using aggregated date of collective behavior to be "a good way to meet people" (PEW, 2015). Though, the question wasn't exactly worded so concretely. We're already happy to give up some of our autonomy, and may be happy to give up even more as new technologies demonstrate their value. Humans generally are terrible at recognizing slow aggreged changes. This makes it difficult to predicted how much autonomy we as a species might be willing to cede to new technologies. For instance, I could imagine a small bodily sensor the reads my current biochemical state and predicts what type of food would be most satisfying at that moment. Then orders it for delivery on my behalf. This tool would leave me with only one unbalanced choice: "Should I eat what I know will be enjoyable or laboriously search for something else?" Perhaps the very task of choosing your meal could improve the eating experience. No worries, there is a large body of literature on how to create the illusion of choice and improve the quality of life within confined systems. Each step along the long path towards our eventual loss of all autonomy could seem rather justified at the time. However, even if justified, the idea of giving up an ever-increasing amount of autonomy is an uncomfortable proposition.

Cybernetic integration does not necessarily mean the loss of our individual self. The Matrix movies explore a world where humans have all been cybernetically connected to a new virtual Earth as the real one is managed by enslaving robots. The individual still exists, but we are all part of one large collective network; kept docile by our shared virtual reality experience. The virtual world is entirely real to the individuals that inhabit it. Each person has the sense of autonomy, but their lives are entirely fabricated by the Matrix, which uses the human bodies as an energy source. I'll leave enslaving robots for a different book, but the idea that we could integrate into a single superorganism without losing our individuality is an intriguing proposition. The human body is full of symbiotic organism that retain much of their individual character. Colonies of bacteria live in our stomach and gut, which assist with food digestion. These bacteria are fundamental to our survival, but still maintain life form autonomy and their own reproductive cycles. Similarly, most human cells contain mitochondria which maintains its own independent DNA. Tracking mitochondrial DNA, which doesn't change much from generation to generation, is one of the best genetic methods for tracking family heritage. Mitochondria also have a semi-autonomous replication process, despite being integral to our own cellular life cycles. This brings us back to the starship example and re-emphasizes the difficulty in trying to define what a human superorganism might look like. Individuals could participate within a larger system but maintain some level of individuality.

All our examples so far explore the possibilities of cybernetic integration. But science fiction hasn't been limited to human technology. Biological or some yet unknown form of integration may also be possible. My food supply is already fully dependent on the collective free market. Delivering nutrients from one part of the system to the other is the goal of both a circulatory system and a food supply chain. But a food supply chain doesn't necessarily represent the movement towards a superorganism. It is simply another example of humans continuing the fight against entropy by discovering ways to increase complexity and deliver energy savings. Unlike science fiction, this is something we can study and improve.

So, what does the science tell us about the possibilities of a human superorganism? Well...almost nothing. We can theorize based on truly informed research, but those theories are not proven to be predictive. Humans currently live in an energy rich environment, if not an unevenly distributed one. Our path forward is not limited by energy reserves, so much as it is by the side effects of energy extraction and the limits of energy storage. We could collectively choose a science fiction future to work towards, but there are inevitably future challenges we cannot predict. Those future challenges will dictate the course of our collective path. Leaving the future of the human superorganism to the realm of science fiction for now. However, the current state of the path we chose to travel is within our sphere of influence.

There are several scientific theories about the nature of systems which humans participate within, and the future course those systems may take. The technological singularity theory is probably closest to the Star Trek Borg example we explored above. This is the idea that machine learning will eventually surpass human intelligence. Machine learning is the ability of the algorithms running our computers to improve themselves without humans programming each new step forward. People who advocate for the singularity believe that eventually this computer self-improvement will outpace the human ability to understand the improvements. At which point, computers will be writing their own algorithms so complex that they will effectively be smarter than humans. This is likely to happen in different fields of intelligence at different times. While technological singularly is not a proven theory, it is grounded in keen observations of current trends in computing development.

Technologists like Ray Kurzweil predict this type of computing intelligent could create itself as early as

2050. Other technologists argue that the entire theory is closer to science fiction. However, most naysayers generally contest the scope of the singularity, not the possibility that computers are likely to overtake humans in at least a few domains of intelligence. Their ability to solve computational problems has already far surpassed our own, even before algorithms were tasked with improving themselves. So, while the fate of a technological singularity is still unknown, the evidence is strong enough to cause intellectual luminaries like Stephen Hawking and Bill Gates to sound the alarm about a dangerous future for artificial intelligence. The technological singularity has become a catalyst for talking about the disruptive changes artificial intelligence will inevitably cause to human culture, both economically and socially. Most immediately is the impending upheaval driverless cars are likely to bring within the next few years. Much has already been written about this topic, from less cars being manufactured to parking lots disappearing, so I'll focus only on the impending job disruption for drivers. Harvard labor economist Lawrence Katz estimates that 5 million Americans, roughly 3% of the workforce, could lose their jobs in the next few years as self-learning algorithms improve themselves beyond the skill of human drivers.[47] While

---

47 Steven Greenhouse (2016). "Autonomous vehicles could cost America 5 million jobs. What should we do about it?" LA Times

the safety benefits of driverless cars is argument enough to advocate for them, rapidly losing 5 million labor jobs will be a huge disruption to the structure of our economic system. And that is just from one type of algorithm becoming better at driving then we are.

Philosopher David Chalmers argued for a less miserable possibility in his paper "The Singularity: A Philosophical Analysis" (2010). This paper is a great examination of the likelihood of a technological singularity, but his argument for human and A.I. integration is most relevant to our discussion. He describes three possible outcomes where Artificial Intelligence passes us by: extinction, isolation, and inferiority. Each are non-ideal outcomes where humans fall on the losing end of obsolete. Which is why Chalmers argued for the forth possibility: integration:

> In the long run, if we are to match the speed and capacity of nonbiological systems, we will probably have to dispense with our biological core entirely. This might happen through a gradual process through which parts of our brain are replaced over time, or it happen through a process of scanning our brains and loading the result into a computer, and then enhancing the resulting processes. Either way, the result is likely to be an enhanced nonbiological system, most likely a computational system." (Chalmers, 2010)

Chalmers goes on to explain the known barriers to cybernetic integration with human consciousness. However, we do not need to know how to fully transfer human consciousness into a machine before we can begin the work of making sure that artificial intelligence is improving human potential and not simply passing us by. Today's rehabilitation laboratories have already developed amazing integrations between the human brain and artificial limbs. Electronic hands that not only receive movement commands from our brain, but return useful sensory information. Integrating electronic components into mammalian neuronal networks is no longer in the realm of science fiction. Technologies in development today will generate huge changes in the systems which shape human cultures. We need to start listening to the people who study these changes and develop collective goals for how we want to proceed along our path of ever increasing connectivity and cultural complexity.

We cannot know what a human superorganism will look or act like, but knowing the final state is not how superorganisms arise. What we can measure and know is the state of systems that currently shape our cultures. We choose every day to participate in various systems. From our education and work, to global trade and the internet. These systems have structures that emphasize certain human behaviors. For example, prominent Silicon Valley ethicist Tristan Harris warns that many of the social media algorithms we encounter every day, are optimized to elicit our outrage. This is not intentional. It is the result of social media sites needing to keep their users coming back and staying engaged. The algorithms that pick which advertisement or news story you are most likely to engage with, have established that outrage is the best target for your attention. This is not true for everyone, but our collective behavior is what's important here. Harris warns about letting algorithms choose our emotional states and encourages us to become more proactive in determining what these algorithms are programed to optimize. Not through force, but through careful study of how we can incentivize social media companies to prioritize human happiness and wellbeing.

Very few systems ever transition into a superorganism, so there should be no assumption that humans will make this change. Our species may just continue to evolve as autonomous individuals. A lot more researcher is needed before we will understand why some systems transition into a superorganism and others do not. What we do know is the human species will continue fighting the entropy battle. Our social and biological complexity is the result of negentropy. All evidence suggests we will continue our ever-increasing efforts to collect and efficiently consume energy. Collective orchestration within systems is the foundation for increasing that complexity and improving our energy efficiency. We may not know the outcome, but humans have the tools necessary for steering a course based on the knowledge that delaying entropy is our ultimate challenge. We need to be reviewing the systems that currently drive our complexity forward, discovering their likely outcome, and designing a shared environment that encourages as many people to freely participate as possible. This process will provide us with the most resources possible for surviving the unknown future. This is the process of collective orchestration.